Engineering Rock Mechanics

Engineering Rock Mechanics

Contributors

Meng Wang, Zheming Zhu et al.

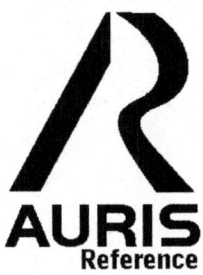

www.aurisreference.com

Engineering Rock Mechanics

Contributors: Meng Wang, Zheming Zhu et al.

Published by Auris Reference Limited

www.aurisreference.com

United Kingdom

Engineering Rock Mechanics

ISBN: 978-1-78154-905-6

British Library Cataloguing in Publication Data
A CIP record for this book is available from the British Library

Printed in the United Kingdom

Exclusively distributed by CBS Publishers & Distributors Pvt. Ltd.

Sales & Distribution Rights only for India, Pakistan, Bangladesh, Sri Lanka, Nepal and Bhutan. This book is not to be sold outside these territories.

Contents

List of Abbreviations

AECL	Atomic Energy of Canada Limited
BDT	Brazilian Disk Test
CB	Chevron Bend
CCNBD	Chevron-Notched Brazilian Disk
CDG	Completely Decomposed Granite
CI	Crack Initiation
COD	Crack-Opening Displacement
CSTBD	Cracked Straight-Through Brazilian Disk
DEM	Distinct Element Method
DI	Depth Index
DPI	Drill Parameter Interpretation
FEM	Finite Element Method
FFRC	Free–Free Resonant Column
FPZ	Fracture Process Zone
GCD	Gouge Content Designation
HDG	Highly Decomposed Granite
ISRM	International Society for Rock Mechanics
LPD	Load-Point Displacement
LPI	Lithology Permeability Index
MTS	Maximum Tangential Stress
MWD	Measurement While Drilling
NGI	Norwegian Geotechnical Institute
REV	Representative Elementary Volume
RQD	Rock Quality Designation
RQD	Rock Quality Designation
SCB	Semi-Circular Bend
SNDB	Straight Notched Disk Bending
UCS	Uniaxial Compressive Strengths
XRD	X-Ray Diffractometry

List of Contributors

Meng Wang
Department of Engineering Mechanics, Sichuan University, Chengdu 610065, China

Zheming Zhu
Department of Engineering Mechanics, Sichuan University, Chengdu 610065, China

Jun Xie
Department of Engineering Mechanics, Sichuan University, Chengdu 610065, China

Yifeng Chen
State Key Laboratory of Water Resources and Hydropower Engineering Science, Key Laboratory of Rock Mechanics in Hydraulic Structural Engineering, Wuhan University, P. R. China

Chuangbing Zhou
State Key Laboratory of Water Resources and Hydropower Engineering Science, Key Laboratory of Rock Mechanics in Hydraulic Structural Engineering, Wuhan University, P. R. China

Are Håvard Høien
Norwegian Public Roads Administration, Postboks 8142 Dep, 0033 Oslo, Norway

Bjørn Nilsen
Norwegian Public Roads Administration, Postboks 8142 Dep, 0033 Oslo, Norway

Diyuan Li
School of Resources and Safety Engineering, Central South University, Changsha 410083, Hunan, China

Charlie C. Li
Department of Geology and Mineral Resources Engineering, The Norwegian University of Science and Technology (NTNU), 7491 Trondheim, Norway

Xibing Li
School of Resources and Safety Engineering, Central South University, Changsha 410083, Hunan, China

S. R. Hencher
Halcrow China Ltd., Hong Kong, China

S. G. Lee
University of Seoul, Seoul, Korea

T. G. Carter
Golder Associates, Toronto, Canada

L. R. Richards
Canterbury, New Zealand

Takahiro Funatsu
National Institute of Advanced Industrial Science and Technology (AIST), Central 7, 1-1-1 Higashi, Tsukuba, Ibaraki 305-8567, Japan

Norikazu Shimizu
Yamaguchi University, 2-16-1 Tokiwadai, Ube, Yamaguchi 755-8611, Japan

Mahinda Kuruppu
Curtin University, Locked Bag 30, Kalgoorlie, WA 6433, Australia

Kikuo Matsui
Kyushu University, 744 Motooka, Nishi-ku, Fukuoka 819-0395, Japan

Cheng-Yu Ku
National Taiwan Ocean University

Shih-Meng Hsu
Sinotech Engineering Consultants, Inc Taiwan

Tae-Min Oh
Department of Civil and Environmental Engineering, Korea Advanced Institute of Science and Technology (KAIST), Taejon 305-701, Korea

Tae-Hyuk Kwon
Earth Sciences Division, Lawrence Berkeley National Laboratory, 1 Cyclotron Rd. MS 90R1116, Berkeley CA 94720, USA

Gye-Chun Cho
Department of Civil and Environmental Engineering, Korea Advanced Institute of Science and Technology (KAIST), Taejon 305-701, Korea

Preface

Engineering rock mechanics is the discipline used to design structures built in rock. These structures encompass building foundations, dams, slopes, shafts, tunnels, caverns, hydroelectric schemes, mines, radioactive waste repositories and geothermal energy projects: in short, any structure built on or in a rock mass. The text Engineering Rock Mechanics covers the basic rock mechanics principles. First chapter presents a comparison of the simulation of uniaxial compressive damage and true triaxial unloading failure by using a combination of experimental and numerical simulation. Second chapter focuses on stress/strain-dependent properties of hydraulic conductivity for fractured rocks. In third chapter, the relationships between the drill parameter interpretation (DPI) factors water and fracturing are examined in relation to grout volumes. The purpose of fourth chapter is to find out the condition to create slabbing failure under uniaxial compression and to determine the slabbing strength of hard rock in the laboratory. Fifth chapter reviews several landslide case histories and provides guidelines for characterizing sheeting joints and determining their shear strength. In sixth chapter, we investigate the mode I fracture toughness using a semi-circular bend (SCB) specimen. Seventh chapter presents the measured hydraulic conductivity results and the relationship among the hydraulic conductivity, RQD, DI, GCD, and LPI. Effect of partial water saturation on attenuation characteristics of low porosity rocks has been investigated in last chapter.

Chapter 1

AN EXPERIMENTAL STUDY ON DEFORMATION FRACTURES OF FISSURED ROCK AROUND TUNNELS IN TRUE TRIAXIAL UNLOADS

Meng Wang, Zheming Zhu, and Jun Xie

Department of Engineering Mechanics, Sichuan University, Chengdu 610065, China

ABSTRACT

Joints and cracks are frequently encountered in underground rock mass. During the process of tunnel excavations or other underground construction, the rock will be exposed suddenly, and such sudden unloading process will cause crack expansion and destabilize the rock structure. In order to investigate the crack behaviour during this process, a true triaxial loading apparatus with a computer-controlled electrohydraulic servosystem was established, and a series of true triaxial loading and unloading experiments was conducted by using concrete specimens containing inclined cracks with inclinations of 15°, 30°, 45°, 60°, and 75°. The stress-strain behavior and the failure property of rock models during unloading process were obtained, and, additionally, the coefficient of brittle stress drop was investigated. The uniaxial compression tests were simulated by using finite element method.

INTRODUCTION

Tunnels constructed for road traffic and mineral mining in China can reach depths of 1000–2000 m. Deep roadway excavation usually results in unidirectional or two-direction unloading, which can weaken the stability of the surrounding rocks, leading to rock damage, typically unloading damage [1, 2]. Deep roadway excavation is a high-stress unloading process and may cause severe expansion of the rock in the unloading direction. The damage occurs mainly as tension fractures, as well as tensional shear fracture and shear fracture [3]. The failure characteristics of basalt, granite, and sandstone under unloading conditions have been studied extensively [4–7]; D. Huang and R. Q. Huang [8] examined the brittle stress drop of granite under equal triaxial confining pressure. Research on the related unloading failure mechanism based on damage-fracture mechanics is in progress to establish a complete stress-strain model under unloading conditions, including the nonlinear strengthening stage, stress drop stage, and strain softening stage [9].

Underground rock mass is subjected to three-direction loads, and they are not the same due to many factors involved, such as the geological structure, the nearby volcanic eruptions, or earthquakes. Joints and cracks are frequently encountered in the underground rock mass, and such cracks usually play a dominant role in the stability of brittle material structures. During the process of rock excavation, the cracks will be exposed and the state of stresses to which the cracks are subjected will alter, and subsequently, the cracks may propagate and may lead to disasters, such as rock burst. Therefore, it is necessary to study the behaviors of cracks during the unloading process.

Zhu et al. [10–15] conducted a number of studies on fracture criteria for a single crack and collinear cracks under compression loading conditions, but the crack behavior during the unloading process has not been focused. D. Huang and R. Q. Huang [8] conducted experiments to study the evolution of central fracture transformation and extension under equal confining pressure and unloading conditions. Xu and Jiang [16] discussed rock transformation and destruction in different loading stress paths based on a true triaxial experiment, and simulations of rock multistress path evolution under high ground stress were performed by Chen and Feng [17]. True triaxial experiments can simulate several stress paths and represent the complicated underground working environment of a mine. Underground excavation is dangerous because of the existence of cracks in the surrounding rock. Based on the photoelastic experimental method, Wang et al. [18] investigated the stress intensity factors of cracks surrounding tunnels with variable fracture inclination and discussed the mechanisms of unstable tunnel destruction caused by cracks. However, the unloading behavior of tunnels has not been addressed.

In this study we carried out triaxial unloading failure experiments by using a true triaxial loading device which is a computer-controlled electrohydraulic servo system. A group of physical models with different inclination angle cracks close to the unloading surface was conducted and was loaded by the true triaxial device. The stress-strain characteristics and failure property of the rock models with different inclination angle cracks during unloading process were obtained. Additionally, the coefficient of brittle stress drop based on these physical models was determined.

EXPERIMENT

Preparation of Physical Models

The materials of lime, cement, and silver sand at a ratio of 4 : 7 : 11 were used to construct the physical models. Hardened lime is brittle but has lower surface density and low strength; hardened cement has lower brittleness but high strength. Using cement and lime as gel materials creates a model that is brittle like rock and also of high strength and elasticity. The model size was $150 \times 150 \times 300$ mm, the dry density ρ was about 2750 kg/m^3, and the regular triaxial experiment elasticity E was 2.4 GPa, with a Poisson ratio n of 0.23.

Five different models were created with crack inclinations of 15°, 30°, 45°, 60°, and 75°; each model was made of six materials with a crack length of 50 mm. The spatial relationship is shown in Figure 1. To form the crack, the model material was poured over a 0.2 mm thick rigid plastic film that represented the crack. When the desired shape and strength were obtained, the material was heated and the plastic film was removed. When the material cooled, the crack closed naturally.

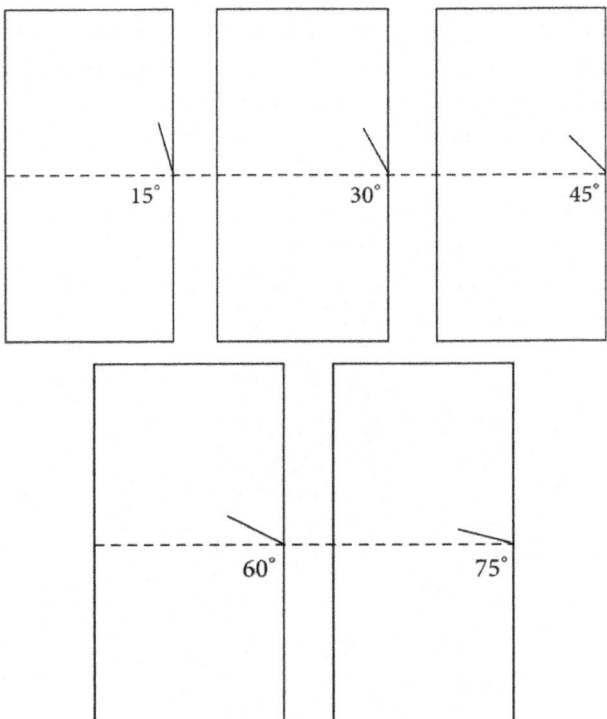

Figure 1: Diagram of the testing models.

The specimens used in the experiment were large; therefore, joints and cracks often occur in the interior of the specimen, affecting the experimental results. Hence, a special processing method was used that is suitable for analyzing naturally closing cracks. The method uses the thermal expansion and contraction properties of the material. A cutting method is used to create the cracks in the rock specimens, which results in a certain distance between the faces of crack; this affects the curve of the vertical axial stress versus strain. The mechanical properties of the specimens are similar to those of rock and the parameters of density and E are very close, as shown above.

Loading/Unloading Path

In the experiment, a true triaxial experimental apparatus was established by using computer-controlled electrohydraulic servo system, which was different from the MTS test machine [19]. The position of the model under load testing is shown in Figure 2. The lateral loads were hydraulic loads. The testing equipment measures and records the surface pressure and displacement of the specimens.

The feasibility of the testing method was first verified under loading and unloading conditions. Loading the specimens causes the cracks to extend and leads to destruction of the specimen. The experiment included two tests, a reference test and an unloading test, to confirm that material crack extension and connection are caused by unloading. The specific plan is as follows.

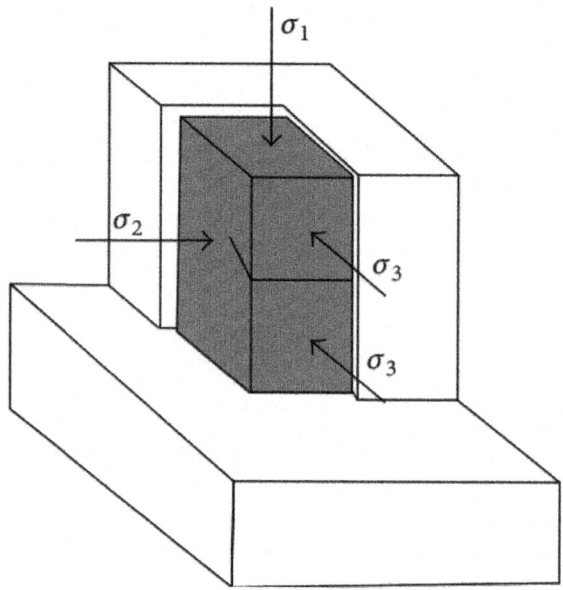

Figure 2: Diagram of load of testing model.

Reference Test

(a) Slowly add triaxial pressure to the designed load (15.6 MPa in the axial direction, 6.5 MPa in the lateral direction, and 5.7 MPa to the unloading surface). The vertical axial compression pressure controls the displacement (at a loading rate of 0.017 mm/s); the confining pressure controls the stress (at a loading rate of 0.03 MPa/s). (b) Maintain these conditions for 30 s. (c) Slowly unload in the axial direction at an unloading rate of 0.017 mm/s. Once reaching zero, unload the triaxial pressure at an unloading rate of 0.03 MPa/s.

Unloading Test

(a) Slowly add triaxial pressure to the designed load (15.6 MPa in the axial direction, 6.5 MPa in the lateral direction, and 5.7 MPa to the unloading surface). The vertical axial compression pressure controls the displacement (at a loading rate of 0.017 mm/s) and the confining pressure controls the stress (at a loading rate of 0.03 MPa/s). (b) Maintain these conditions for 30 s. (c) Quickly unload to the unloading surface at an unloading rate of 2 MPa/s while maintaining all the other conditions at their original state.

Upon loading, increase the pressure on the top surface, the excavated face, and the side of the model while constraining the other sides by normal displacement (Figure 2; σ_3 is the load of the excavated face).

Comparing the two tests for the same crack angle we see that, in the reference test, the materials remain intact after unloading and are in a state of densification, with no crack extension. However, in the unloading test, the materials are destroyed. This indicates that the extension and merging of the cracks are indeed caused by unloading.

NUMERICAL SIMULATION AND ANALYSIS

Since the studies by Griffith [20] showed the propagation of crack instability and the growth criterion introduced by Irwin [21] based on the stress intensity factor, fracture theory has undergone significant development. However, previous studies were limited to simple fracture behavior. Theoretical models that examined the mechanism for crack initiation, expansion, and coalescence were established in recent years with the development of experimental techniques and computer technology. This paper presents a comparison of the simulation of uniaxial compressive damage and true triaxial unloading failure by using a combination of experimental and numerical simulation.

The paper uses the finite element software ANSYS to analyze and simulate crack initiation, expansion, and coalescence under uniaxial compression by using the theory based on the overall failure criterion of rock energy dissipation [22]. For the two-dimensional model, $\sigma_2 = 0$. Thus, the formula found in the literature reduces to the following formula:

$$\sigma_c^3 = (\sigma_1 - \sigma_3)\left(\sigma_1^2 + \sigma_3^2 - 2\nu\sigma_1\sigma_3\right),$$

$$\sigma_t^3 = \sigma_3\left(\sigma_1^2 + \sigma_3^2 - 2\nu\sigma_1\sigma_3\right). \tag{1}$$

The model uses a crack angle of 45° and uniaxial compression as the boundary condition (Figure 3).

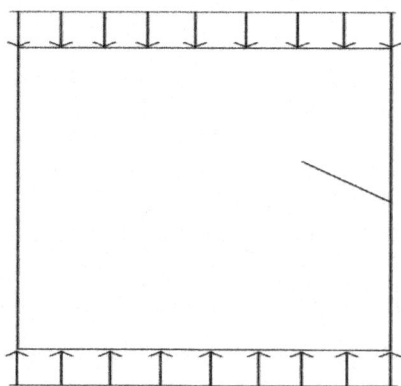

Figure 3: The numerical simulation models.

The numerical results show that the expansion mode of the crack, combined with predictive modeling based on fracture mechanics theory, is similar to that observed in the uniaxial compression test. In particular, the trajectories of the preexpansion mode are close to each other (Figure 4). These results, however, are quite different from the experimental results under triaxial unloading conditions (Figure 5). Under unloading conditions, it is difficult to use the existing and established theories to discuss and analyze the damage.

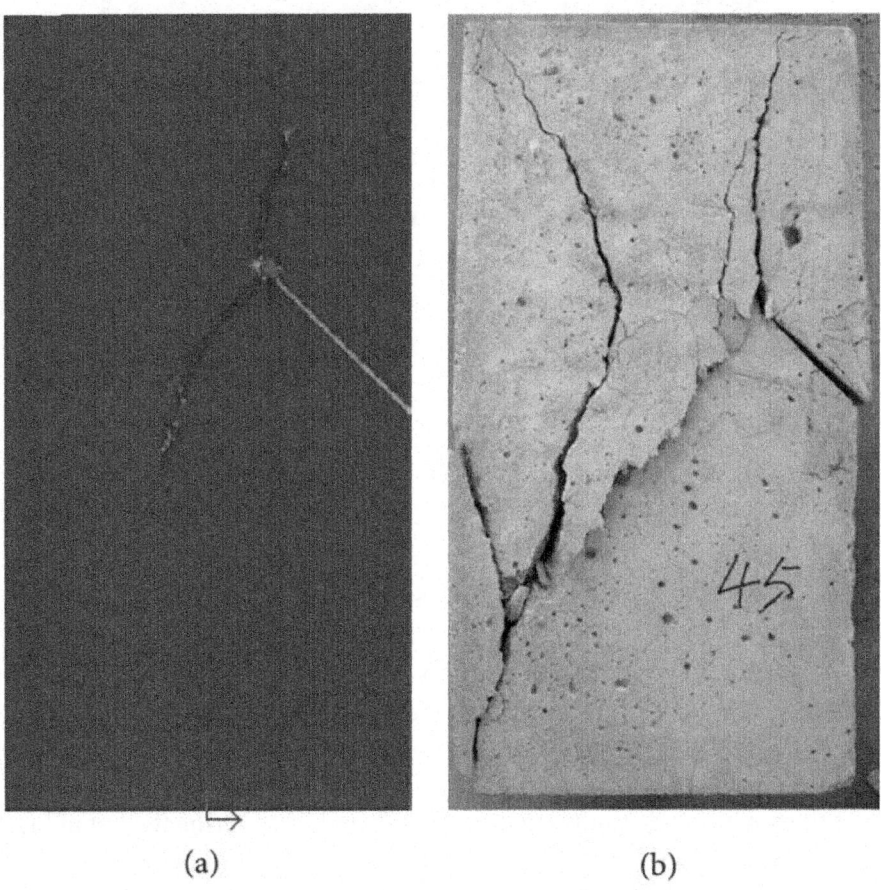

(a)   (b)

Figure 4: Failure of fissured rocks: photos and numerical simulation. Numerical simulation (a) and uniaxial compression test (b).

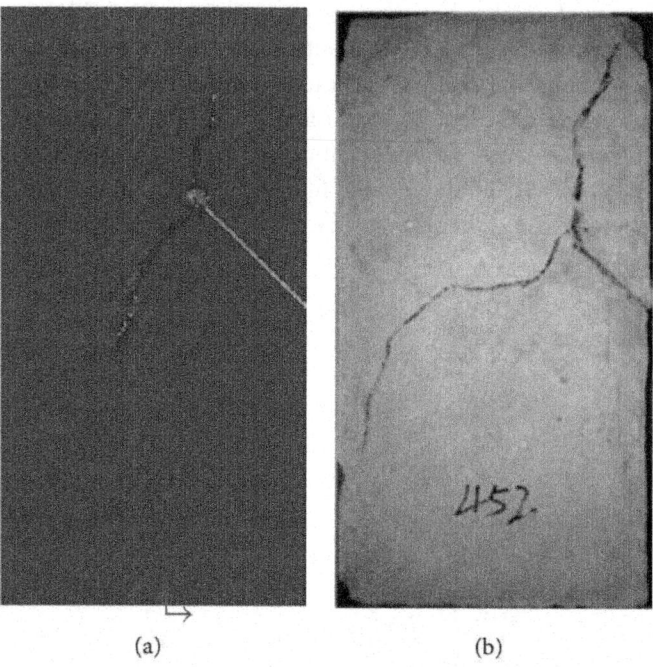

(a)   (b)

Figure 5: Failure of fissured rocks: photos and numerical simulation. Uniaxial compression test (a) and true triaxial unloading test (b).

ANALYSIS OF EXPERIMENT RESULT

Stress-Strain Characteristic Curve under Unloading Condition

The experimental results were processed by the Origin software package to construct the stress-strain curves which were then smoothed. Figure 6 shows typical complete axial-directional stress-strain curves under unloading conditions with crack inclination of 15°, 45°, and 75°. During the initial loading stage, the downward curve (OA) is obvious and presents a typical densification. As the crack angle becomes wider, the axial nominal strain generated when the crack completely closes increases. Therefore, as the crack becomes bigger, the densification becomes more obvious. This process is a reflection of the crack close-up in the loading phase. Before reaching the peak value of the axial-directional stress, the curve shows nonlinear deformation, but no yield platform. Judging from the after-peak curve, in the unloading phase, the axial-directional stress falls quickly to the residual strength level, with a sharp curve slope and small axial-directional deformation. It shows clear characteristics of brittle stress drop. In addition, for the models with a large inclination, the curve shows an abrupt change. When the axial-directional stress falls to residual strength level, the destruction is rapid and produces a large sound.

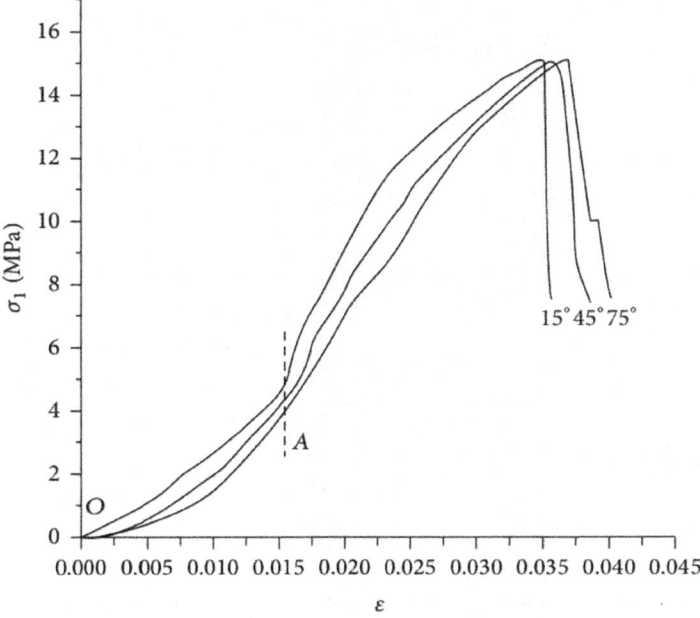

Figure 6: Vertical axial stress versus strain under unloading.

Failure Characteristics

Figure 7 shows a typical failure form and its corresponding sketch. The different crack propagation forms indicate that during the test the crack expansion consists of both tensional shear failure and shear failure. When the crack angle is set at 15°, a pair of tensional fractures appears on the crack tip, with no shear fractures around. During the unloading process, tensional fractures are observed; then shear fractures appear away from the unloading surface. During the entire process, preliminary unloading causes expansion along the unloading direction, and the unloading surface protrudes outward. Meanwhile, in Figure 7, the regional fracture in the red circle connects with the unloading surface, producing a spalling failure. At a later stage, the material generates a perfoliate fracture (besides the original fracture expansion) along the axial direction. When the crack angles are set at 30° and 45°, at the initial stage, tensional failure occurs at the fracture tip and expands upward along the unloading surface, and the fracture expands inward at the bottom. The two fractures connect and create a macro failure. Impacted by unloading, the fracture expansion is parallel to the unloading surface. When the crack angles are set at 60° and 75°, tensional failure appears at the initial stage and expands upward. At the same time, the fracture expands inward at the lower part. The two fractures connect and create a macro failure.

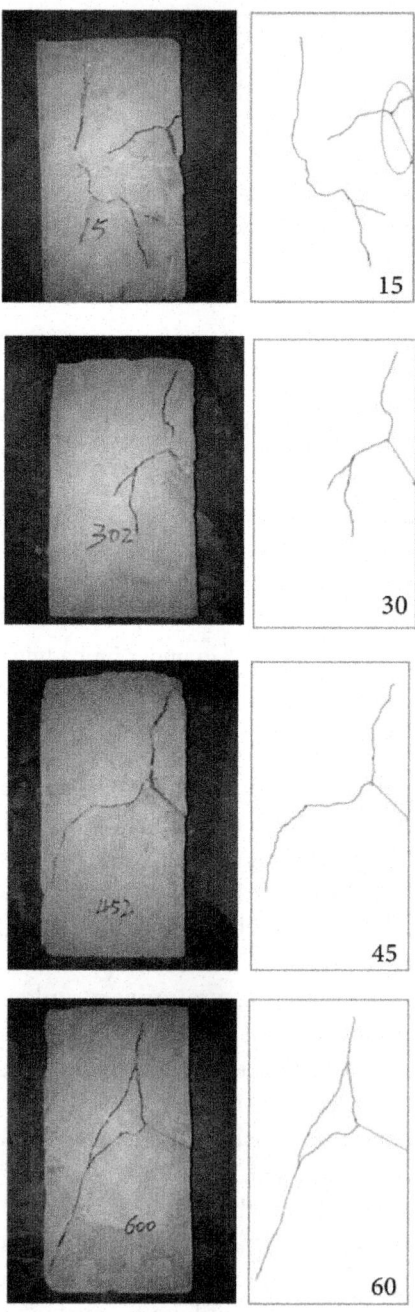

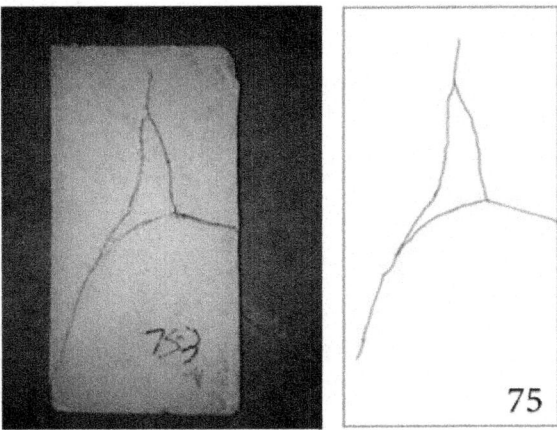

Figure 7: Failure of fissured rocks: photos and diagram.

In conventional materials, the material is considered as ductile material when it has large deformation but is not broken. Otherwise, it is considered as brittle material. Brittleness and ductility are mainly composed of methods of rock disintegration to distinguish [23]. The rupture and yield are seen as brittle failure and ductile failure, and the strain and the profile of damage can be used as the main control standards.

From Figure 6, the strain of the specimen at failure can be observed, and Heard [24] uses 3% and 5% for boundaries to distinguish brittle and ductile rock. In this experiment, the strain changes from 3.5% to 4.1%, and it illustrates the tendency from brittle to ductile transition.

The above results show that the bigger the crack angle, the greater the failure. During the unloading process, fractures generated at the initial stage join; this is especially obvious for the models with crack inclinations of 60° and 75°. Moreover, the direction of the penetrating crack changes from its original direction; as the unloading progresses, the crack's extended fracture tends to align with the axial direction. Thus, during underground excavation, when encountering cracks and joints that are almost parallel to the excavating direction, shotcrete stabilization should be applied to avoid rock spalling rather than using anchors; when encountering cracks and joints that are perpendicular to the excavation direction, anchor bolt support should be applied to prevent larger cracks from developing.

In the relevant reference [25], the zonal disintegration phenomena were clearly showed by unloading the σ_3 stress. As shown, there was a certain distance between the failure surface and the excavation face. The condition of the reference is equivalent to the specimen containing an inclined crack with inclination of 90°. In this paper, the specimen containing an inclined crack with inclination of 75° has shown the zonal disintegration phenomenon. The comparison shows the experiments were significant, and they were right, as shown in Figure8. The result in this paper was shown as Figure 8(a), and the result of the reference was shown as Figure 8(b).

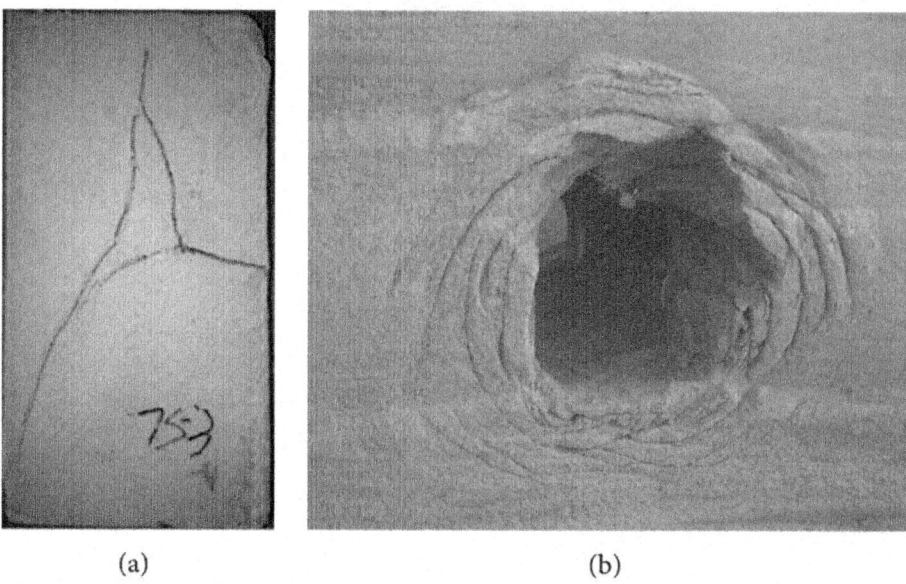

(a)   (b)

Figure 8: The comparison of the results between this paper and the reference.

Comparison Characteristics

Figure 9 shows a comparison of the triaxial failure modes and uniaxial test failure modes. In the uniaxial experiment, both σ_2 and σ_3 are zero. The fractures that lead to failure all pass through the fracture tip, and the overall fracture direction is close to that of the cracked surfaces. The shear effect is significant and causes a serious failure which leads to secondary failure. The triaxial failure is a unidirectional failure caused by dilatation; as the inclination angle increases, delamination failure appears parallel to the unloading surface. In the experiment, the phenomenon is clearly observed for inclinations of 60° and 75°. In unidirectional unloading, because the energy is released in the same direction, rapidly falling rock fragments appear when high energy is released at a small crack angle. Unidirectional unloading is commonly seen in excavation, and there is a considerable difference between the unloading experiment and the single-axis compression failure test, reflecting the significance of true triaxial unloading experimental research.

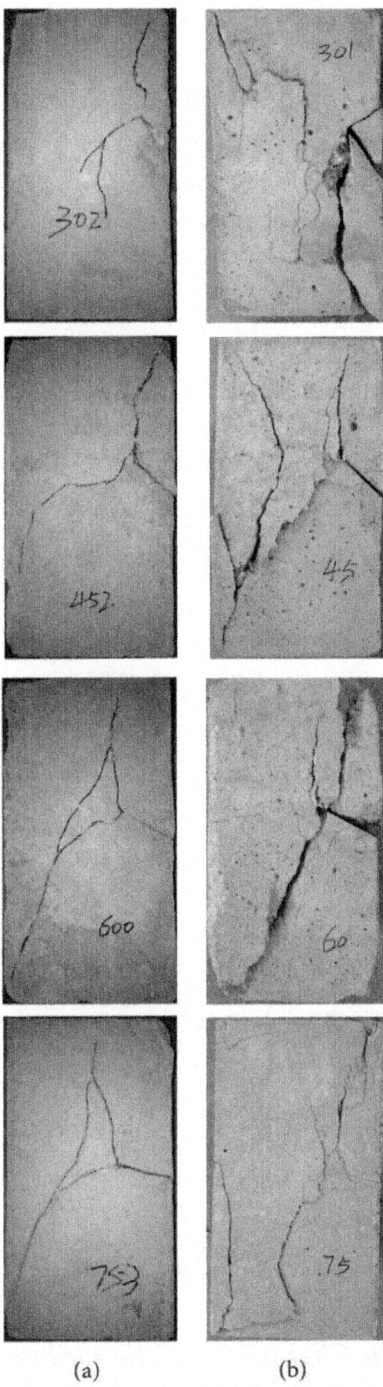

(a)   (b)

Figure 9: Failure of fissured rocks. True triaxial unloading test (a) and uniaxial compression test (b).

Coefficient of Brittle Stress Drop

Parameter Confirmation Method

Using plastic potential theory, Ge [26] derived the brittle stress drop process of the stress nonvertical drop model. Shi et al. [27] combined the overall stress-strain curve of a typical rock compression test with observed brittleness as shown in Figure 10 and propose the coefficient of brittle stress drop R:

$$R = \frac{b}{a},$$

$$(2)$$

where a and b are strain-related parameters, $a = \varepsilon_p - \varepsilon_M$, $b = \varepsilon_B - \varepsilon_p$, and ε_p is the axial strain of peak intensity point; ε_B is the axial strain of the residual intensity point; and ε_M is the difference between the corresponding initial elastic loading strain and the residual intensity stress value. The ideal brittleness model is the special situation, where $b=0$. Equation (2) indicates that the smaller the R is, the stronger the failure character of the rock stress brittleness is.

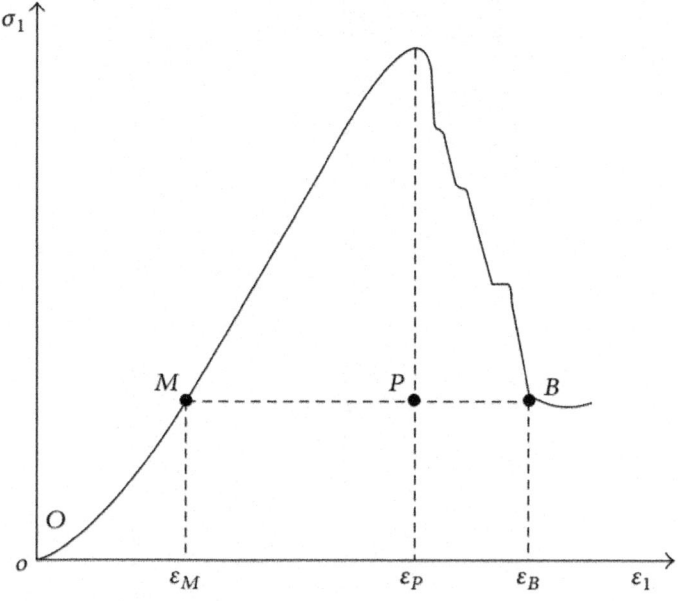

Figure 10: Typical complete stress-strain curve of brittle rocks.

Parameter Variation Characters in the Unloading Test

Studies on the relationship between the coefficient of brittle stress drop of intact rock and the uniaxial/equal triaxial confining pressure and general confining pressure include those of Huang et al. [5] and Chen and Feng [17]. However, during underground excavations, faults and joints will often be encountered, and the coefficient of brittle stress drop will be related to the crack angle. Under the same initial stress and confining pressure, the coefficient of brittle stress drop and the characteristic parameters in the unloading test should be a function of crack angle, as in

$$R(\theta) = \frac{\varepsilon_B(\theta) - \varepsilon_P(\theta)}{\varepsilon_P(\theta) - \varepsilon_M(\theta)}. \tag{3}$$

The axial-directional stress rapidly falls to residual strength levels. This feature is caused by the rapid expansion and connectivity of the cracks. The brittle fracture properties of rock specimens and the damage and fracture mechanism of brittle rock and their strength characteristics were observed by Costamagna et al. [28] and Golshani et al. [25]. However, few studies on the properties of brittle fracture under triaxial unloading conditions have been conducted. The coefficients of brittle stress drop can significantly affect the degree of brittleness of the material and it is conducive to improve the nonideal elastic-brittle-plastic model to account for the coefficients of brittle stress drop. Further, it can provide a better theoretical basis for numerical simulations of rock mechanics.

The characteristic parameters of each unloading test material determined during our experiments are presented in Figures 11–13. Figure 12 shows that as the crack inclination angle increases, the axial strain at the peak intensity point tends to lie more along the axial direction. This is mainly because of the compaction at the initial stage. There are gaps between cracks; when the angle between the crack and the horizontal axis is wider, the projected area along the vertical axis becomes larger, which leads to a larger variation upon closing. Figure11 shows that as the crack inclination angle increases, the residual intensity point of the specimen tends to increase along the axial direction (mainly influenced by the peak intensity point).

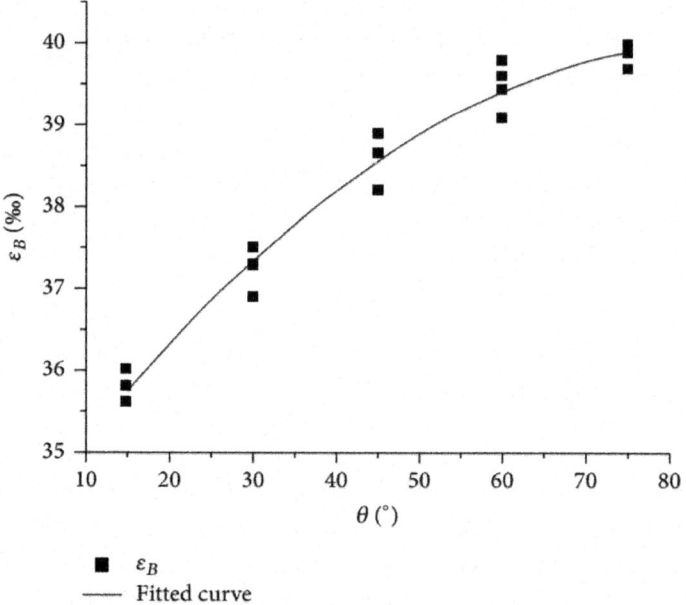

Figure 11: Axial strain at residual strength level versus crack orientations.

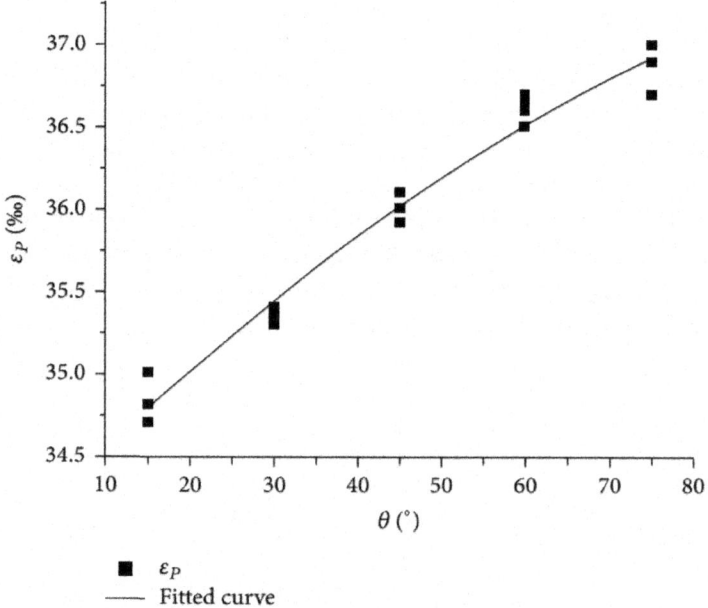

Figure 12: Axial strain at peak strength versus crack orientations.

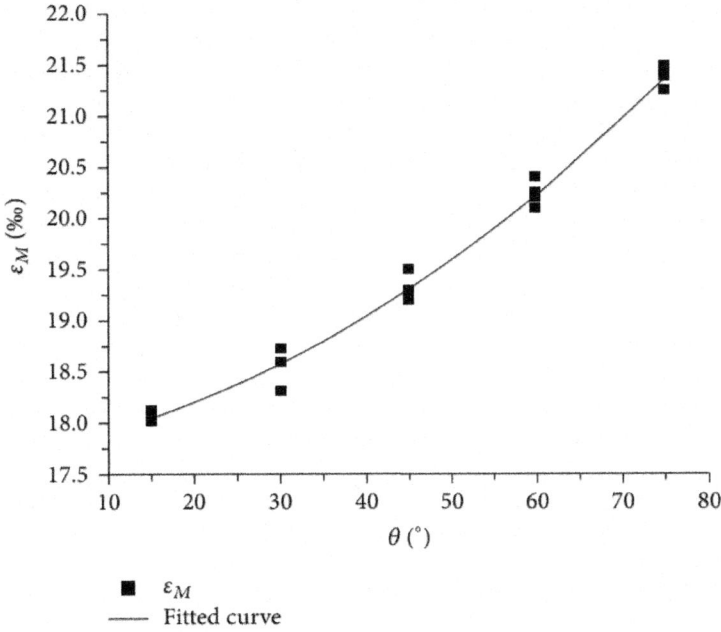

Figure 13: Axial strain at the elastic stage which corresponds to the residual strength versus crack orientations.

Variation of the Brittle Stress Drop Parameters

The author used a quadratic equation to fit the test results (Figures 11–13) to a regression curve, as follows:

$$\varepsilon_B = -0.000820x^2 + 0.143180x + 33.784653,$$

$$\varepsilon_P = -0.000187x^2 + 0.052498x + 34.039389,$$

$$\varepsilon_M = 0.000459x^2 + 0.014482x + 17.718801. \tag{4}$$

Combined with (3) R can be obtained as

$$R(\theta) = \frac{-633 * \theta^2 + 90682 * \theta - 254736}{-646 * \theta^2 + 38016 * \theta + 16320588}. \tag{5}$$

From (5), the curve of the coefficient of brittle stress drop versus crack orientations was derived (Figure 14). As the crack angle widens the brittle stress drop of the specimen increases, suggesting a transformation from brittle to ductile state. Hence, in underground excavation, as the angle between the crack and the unloading surface becomes smaller, the brittle failure becomes more obvious. Simultaneously, brittle failure may suddenly occur. At this point in the excavations, support measures will be required to control the brittle failure. For example, conventional tunnel excavation applies shotcrete-bolt support and anchors penetrate the cracks.

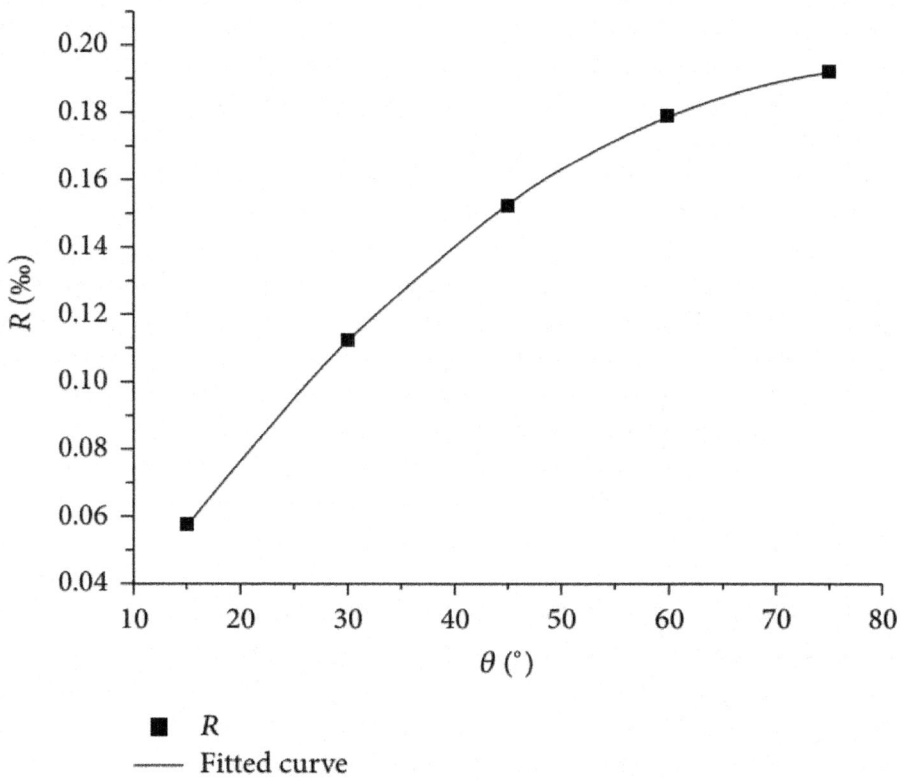

Figure 14: Coefficients of brittle stress drop versus crack orientations.

Deformation during the Unloading Process

In underground excavation, rock deformation is an important factor for project safety evaluations. Based on unloading test results, this paper studies the unloading deformation of the specimen by the axial-directional strain of the after-peak curve.

Figure 15 shows the unloading deformation versus crack orientations. Here, we define the unloading deformation as the difference between ε_p and ε_B.

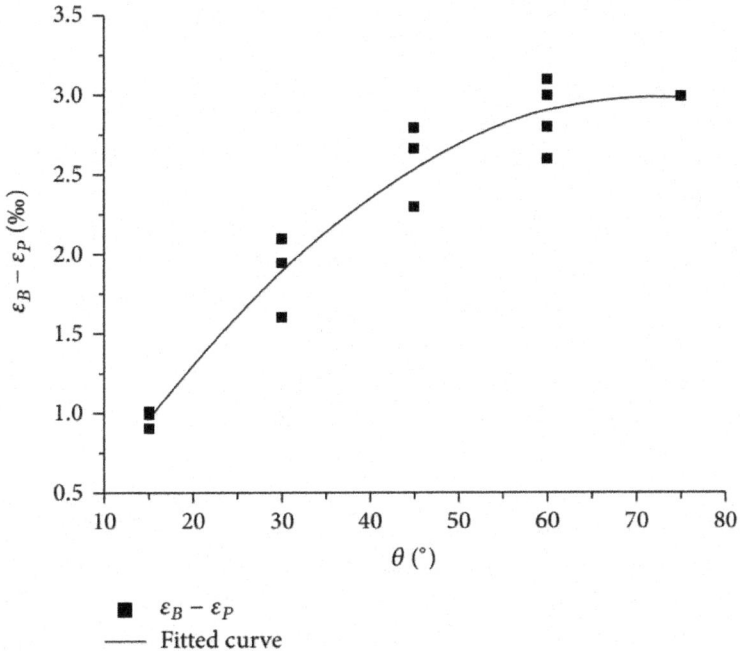

Figure 15: Unloading deformation versus crack orientations.

As the specimen fracture angle increases, the unloading deformation shows an upward trend. The wider the crack angle, the clearer the expansion along the unloading surface. Moreover, the unloading deformation and inclination show good correlation. The quadratic formula can be fitted to this curve to obtain

$$y = -0.0001x^2 + 0.0013x - 0.0003, \qquad (6)$$

$$R^2 = 0.9729. \qquad (7)$$

CONCLUSIONS

In this study, an improved true triaxial experimental facility by introducing a computer-controlled electrohydraulic servo system was established, and triaxial unloading experiments were conducted. A deep-buried tunnel was considered, and the cracks near the unloading surface were investigated. Under unloading conditions, the rock stress-strain characteristics and the failure features for different inclination cracks were studied, and the following conclusions can be obtained.

- During the initial loading stage, the stress-strain curve is concaved and presents typical densification. Before reaching the peak value of the axial-directional stress, the curve shows nonlinear deformation, but no yield platform can be observed.

- From the after-peak stage in the unloading curve, the axial-directional stress falls quickly to the residual strength level with a small axial-directional deformation. It shows the characteristics of brittle stress drop.

- As the specimen crack inclination angle increases, the unloading deformation increases. The larger the crack inclination angle, the clearer the expansion along the unloading surface. The risk of rock collapse during excavation increases as the inclination angle decreases.

- The main form of damage in the failed specimens was tensile failure. The tensile failure characteristics are more obvious in triaxial compression than in uniaxial compression.

- As crack inclination angle increases, the axial nominal strain generated when the crack completely closes increases, and the brittle stress drop increases, and the transformation from brittle to ductile state occurs. Hence, in underground excavation, as the crack inclination angle between the crack and the unloading surface is small, the brittle failure will be more obvious.

- In the further study, the dynamic unloading process should be considered because the dynamic loading [29–31] is different from the static loading.

ACKNOWLEDGMENTS

This work was financially supported by the Open Fund of State Key Laboratory of Oil and Gas Reservoir Geology and Exploitation (PLN1202), by the Project of Science and Technology of Sichuan Province (2014JY0002), and by the National Natural Science Foundation of China (51074109).

REFERENCES

1.   G. Wu, "Current status and prospects of research on mechanism for unloading failure of engineering rock mass," Journal of Engineering Geology, vol. 9, no. 2, pp. 174–181, 2001.

2.   Z. Y. Zhang, S. T. Wang, and L. S. Wang, Principles of Engineering Geology, Geological Publishing House, Beijing, China, 1994.

3.   J. H. Shen, L. S. Wang, Q. H. Wang, J. Xu, Y. Jiang, and B. Sun, "Deformation and fracture features of unloaded rock mass," Chinese Journal of Rock Mechanics and Engineering, vol. 22, no. 12, pp. 2028–2031, 2003.

4. T. B. Li and L. S. Wang, "An experimental study on the deformation and failure features of basalt under unloading condition," Chinese Journal of Rock Mechanics and Engineering, vol. 12, no. 4, pp. 321–327, 1993.

5. D. Huang, R.-Q. Huang, and Y.-X. Zhang, "Characteristics of brittle failure and stress drop under triaxial loading and unloading," Journal of Civil, Architectural & Environmental Engineering, vol. 33, no. 2, pp. 1–6, 2011.

6. J. L. Li, R. H. Wang, Y. Z. Jiang, J. Liu, and X. Chen, "Experimental study of sandstone mechanical properties by unloading triaxial tests," Chinese Journal of Rock Mechanics and Engineering, vol. 29, no. 10, pp. 2034–2041, 2010.

7. Y. H. Lu, Q. S. Liu, and Y. H. Hu, "Damage deformation characteristics and its strength criterion based on unloading experiments of granites," Chinese Journal of Rock Mechanics and Engineering, vol. 28, no. 10, pp. 2096–2103, 2009.

8. D. Huang and R. Q. Huang, "Physical model test on deformation failure and crack propagation evolvement of fissured rocks under unloading," Chinese Journal of Rock Mechanics and Engineering, vol. 29, no. 3, pp. 502–512, 2010.

9. X. P. Zhou, Q. L. Ha, Y. X. Zhang, J. H. Wang, and K. S. Zhu, "Analysis of localization of deformation and complete stress-strain relation for mesoscopic heterogenous brittle rock materials when axial stress is held constant while lateral confinement is reduced," Chinese Journal of Rock Mechanics and Engineering, vol. 24, no. 18, pp. 3236–3245, 2005.

10. Z. M. Zhu, "New biaxial failure criterion for brittle materials in compression," Journal of Engineering Mechanics, vol. 125, no. 11, pp. 1251–1258, 1999.

11. Z. M. Zhu, "An alternative form of propagation criterion for two collinear cracks under compression,"Mathematics and Mechanics of Solids, vol. 14, no. 8, pp. 727–746, 2009.

12. Z. M. Zhu, "Evaluation of the range of horizontal stresses in the earth›s upper crust by using a collinear crack model," Journal of Applied Geophysics, vol. 88, pp. 114–121, 2013.

13. Z. M. Zhu, L. G. Wang, B. Mohanty, and C. Huang, "Stress intensity factor for a cracked specimen under compression," Engineering Fracture Mechanics, vol. 73, no. 4, pp. 482–489, 2006.

14. Z. M. Zhu, H. P. Xie, and S. Ji, "The mixed boundary problems for a mixed mode crack in a finite plate,"Engineering Fracture Mechanics, vol. 56, no. 5, pp. 647–655, 1997.

15. T. Zheng, Z. Zhu, B. Wang, and L. Zeng, "Stress intensity factor for an infinite plane containing three collinear cracks under compression," Journal of Applied Mathematics and Mechanics, vol. 94, no. 10, pp. 853–861, 2014.

16. D. J. Xu and N. G. Jiang, "The various stress paths causing deformation and failure in rocks," Rock and Soil Mechanics, vol. 7, no. 2, pp. 17–25, 1986.

17. J. T. Chen and X. T. Feng, "True triaxial experimental study on rock with high geo-stress," Chinese Journal of Rock Mechanics and Engineering, vol. 25, no. 8, pp. 1537–1543, 2006. ·

18. M. Wang, Z. M. Zhu, and L. G. Zeng, "Experimental study on the stress intensity factors of cracks around tunnels," Journal of Sichuan University (Engineering Science Edition), vol. 44, pp. 99–103, 2012.

19. R. G. Deng, X. M. Fu, and J. Xu, "Study of measuring system for triaxial stresses and pore content of rock with MTS uniaxial machine," Chinese Journal of Rock Mechanics and Engineering, vol. 21, no. 6, pp. 892–896, 2002.

20. A. A. Griffith, "The phenomena of rupture and flow in solids," Philosophical Transactions of the Royal Society of London Series A, vol. 221, pp. 163–198, 1920.

21. G. R. Irwin, "Crack extension force for a part-through crack in a plate," Journal of Applied Mechanics, vol. 29, no. 4, pp. 651–654, 1962.

22. H.-P. Xie, Y. Ju, and L.-Y. Li, "Criteria for strength and structural failure of rocks based on energy dissipation and energy release principles," Chinese Journal of Rock Mechanics and Engineering, vol. 24, no. 17, pp. 3003–3010, 2005.

23. S. Z. Wang, "Brittle-ductile transition and plastic-flow networks in rocks," Progress in Geophysics, vol. 8, no. 4, pp. 25–37, 1993.

24. H. C. Heard, "Effect of large changes in strain rate in the experimental deformation of Yule marble," The Journal of Geology, vol. 71, no. 2, pp. 162–195, 1963.

25. A. Golshani, M. Oda, Y. Okui, T. Takemura, and E. Munkhtogoo, "Numerical simulation of the excavation damaged zone around an opening in brittle rock," International Journal of Rock Mechanics and Mining Sciences, vol. 44, no. 6, pp. 835–845, 2007.

26. X. R. Ge, "Post failure behavior and a brittle plastic model of brittle rock," in Computer Methods and Advances in Geomechanics, pp. 151–160, A.A. Balkema, Rotterdam, The Netherlands, 1997.

An Experimental Study on Deformation Fractures of Fissured Rock... 23

27. G. C. Shi, X. R. Ge, and Y. D. Lu, "Experimental study on coefficients of brittle stress drop of marble,"Chinese Journal of Rock Mechanics and Engineering, vol. 25, no. 8, pp. 1625–1631, 2006.

28. R. Costamagna, J. Renner, and O. T. Bruhns, "Relationship between fracture and friction for brittle rocks," Mechanics of Materials, vol. 39, no. 4, pp. 291–301, 2007.

29. Z. M. Zhu, B. Mohanty, and H. Xie, "Numerical investigation of blasting-induced crack initiation and propagation in rocks," International Journal of Rock Mechanics and Mining Sciences, vol. 44, no. 3, pp. 412–424, 2007.

30. Z. M. Zhu, H. Xie, and B. Mohanty, "Numerical investigation of blasting-induced damage in cylindrical rocks," International Journal of Rock Mechanics and Mining Sciences, vol. 45, no. 2, pp. 111–121, 2008. ·

31. Z. M. Zhu, "Numerical prediction of crater blasting and bench blasting," International Journal of Rock Mechanics and Mining Sciences, vol. 46, no. 6, pp. 1088–1096, 2009.

Chapter 2

STRESS/STRAIN-DEPENDENT PROPERTIES OF HYDRAULIC CONDUCTIVITY FOR FRACTURED ROCKS

Yifeng Chen and Chuangbing Zhou

State Key Laboratory of Water Resources and Hydropower Engineering Science, Key Laboratory of Rock Mechanics in Hydraulic Structural Engineering, Wuhan University, P. R. China

INTRODUCTION

In the last two decades there has seen an increasing interest in the coupling analysis between fluid flow and stress/deformation in fractured rocks, mainly due to the modeling requirements for design and performance assessment of underground radioactive waste repositories, natural gas/oil recovery, seepage flow through dam foundations, reservoir induced earthquakes, etc. Characterization of hydraulic conductivity for fractured rock masses, however, is one of the most challenging problems that are faced by geotechnical engineers. This difficulty largely comes from the fact that rock is a heterogeneous geological material that contains various natural fractures of different scales (Jing, 2003). When engineering works are constructed on or in a rock mass, deformation of both the fractures and intact rock will usually occur as a result of the stress changes. Due to the stiffer rock matrix, most deformation occurs in the fractures, in the form of normal and shear displacement. As a result, the existing fractures may close, open, grow and new fractures may be induced, which in turn changes the structure of the rock mass concerned and alters its fluid flow behaviors and properties. Therefore, the fractures often play a dominant role in understanding the flow-stress/deformation coupling behavior of a rock system, and their mechanical and hydraulic properties have to be properly established (Jing, 2003).

Traditionally, fluid flow through rock fractures has been described by the cubic law, which follows the assumption that the fractures consist of two smooth parallel plates. Real rock fractures, however, have rough walls, variable aperture and asperity areas where the two opposing surfaces of the fracture walls are in contact with each other (Olsson & Barton, 2001). To simplify the problem, a single, average value (or together with its stochastic characteristics) is commonly used to describe the mechanical aperture of an individual fracture. A great amount of work (Lomize, 1951; Louis, 1971; Patir & Cheng, 1978; Barton et al., 1985; Zhou & Xiong, 1996) has been done to find an equivalent, smooth wall hydraulic aperture out of the real mechanical aperture such that when Darcy's law or its modified version is applied, the equivalent smooth fracture yields the same water conducting capacity with its original rough fracture. It is worth noting that clear distinction manifests between the geometrically measured mechanical aperture (denoted by b in the context) and the theoretical smooth wall hydraulic aperture (denoted by b^*), and the former is usually larger in magnitude than the latter due to the roughness of and filling materials in rock fractures (Olsson & Barton, 2001).

The ubiquity of fractures significantly complicates the flow behavior in a discontinuous rock mass. The primary problem here is how to model the flow system and how to determine its corresponding hydraulic properties for flow analysis. Theoretically, the representative elementary volume (REV) of a rock mass can serve as a criterion for selecting a reasonable hydromechanical model. This statement relates to the fact that REV is a fundamental concept that bridges the micro-macro, discrete-continuous and stochastic-determinate behaviors of the fractured rock mass and reflects the size effect of its hydraulic and mechanical properties. The REV size for the hydraulic or mechanical behavior is a macroscopic measurement for which the fractured medium can be seen as a continuum. It is defined as the size beyond which the rock mass includes a large enough population of fractures and the properties (such as hydraulic conductivity tensor and elastic compliance tensor) basically remain the same (Bear, 1972; Min & Jing, 2003; Zhou & Yu, 1999; Wang & Kulatilake, 2002). Owing to high heterogeneity of fractured rock masses, however, the REV can be very large or in some situations may not exist. If the REV does not exist, or is larger than the scale of the flow region of interest, it is no longer appropriate to use the equivalent continuum approach. Instead, the discrete fracture flow approach may be applied to investigate and capture the hydraulic behavior of the fractured rock masses. However, due to the limited available information on fracture geometry and their connectivity, it is not a trivial task to make a detailed flow path model. Thus, in practice, the equivalent continuum model is still the primary choice to approximate the hydraulic behavior of discontinuous rocks. The hydraulic conductivity tensor is a fundamental quantity to characterizing the hydromechanical behavior of a fractured rock. Various techniques have been proposed to quantify the hydraulic conductivity tensor, based on results from field tests, numerical simulations, and back analysis techniques, etc. Earlier investigations focused on using field measurements (e.g. aquifer pumping test or packer test (Hsieh & Neuman, 1985)) to estimate the three-dimensional hydraulic conductivity tensor. This approach, however, is generally time-consuming, expensive and needs well controlled experimental conditions. Numerical and analytical methods are also used to estimate the hydraulic properties of complex rock masses due to its flexibility in handling variations of fracture system geometry and ranges of material properties for sensitivity or uncertainty estimations. In the literature, both the equivalent continuum approach (Snow, 1969; Long et al., 1982; Oda, 1985; Oda, 1986; Liu et al., 1999; Chen et al., 2007; Zhou et al., 2008) and the discrete approach (Wang & Kulatilake, 2002; Min et al., 2004) are widely applied. In this chapter, however, only the equivalent continuum approach is focused for its capability of representing the overall behavior of fractured rock masses at large scales.

Among many others, Snow (1969) developed a mathematical expression for the permeability tensor of a single fracture of arbitrary orientation and aperture and considered that the permeability tensor for a network of such fractures can be formed by adding the respective components of the permeability tensors for each individual fracture. Oda (1985, 1986) formulated the permeability tensor of rock masses based on the geometrical statistics of related fractures. Liu et al. (1999) proposed an analytical solution that links changes in effective porosity and hydraulic conductivity to the redistribution of stresses and strains in disturbed rock masses. Zhou et al. (2008) suggested an analytical model to determine the permeability tensor for fractured rock masses based on the superposition principle of liquid dissipation energy. Although slight discrepancy exists between the permeability tensor and the hydraulic conductivity tensor (the former is an intrinsic property determined by fracture geometry of the rock mass, while the latter also considers the effects of fluid viscosity and gravity), when taking into account the flow-stress coupling effect, the above models presented, respectively, by Snow (1969), Oda (1985) and Zhou et al. (2008) were proved to be functionally equivalent for a certain fluid (Zhou et al., 2008). A common limitation with the above models lies in the fact that the hydraulic conductivity tensor of a fractured rock mass is all formulated to be either stress-dependent or elastic strain-dependent. Consequently, material nonlinearity and post-peak dilatancy are not considered in the formulation of the hydraulic conductivity tensor for disturbed rock masses. To address this problem, Chen et al. (2007) extended the above work and proposed a numerical model to establish the hydraulic conductivity for fractured rock masses under complex loading conditions.

Based on the observation that natural fractures in a rock mass are most often clustered in certain critical orientations resulting from their geological modes and history of formation (Jing, 2003), characterizing the rock mass as an equivalent continuum containing one or multiple sets of planar and parallel fractures with various critical orientations, scales and densities turns out to be a desirable approximation. Starting from this point of view, the deformation patterns of the fracture network can be first characterized by establishing an equivalent elastic or elasto-plastic constitutive model for the homogenized medium. On this basis, a stress-dependent hydraulic conductivity tensor may be formulated for the former for describing the hydraulic behavior of the rock mass at low stress level and with overall elastic response; and a strain-dependent hydraulic conductivity tensor for the latter for demonstrating the influences of material non-linearity and shear dilatancy on the hydraulic properties after post-peak loading. This chapter mainly presents the research results on the stress/strain-dependent hydraulic properties of fractured rock masses under mechanical loading or engineering disturbance achieved by Chen et al. (2006), Zhou et al. (2006), Chen et al. (2007) and Zhou et al. (2008).

The stress-dependent hydraulic conductivity model (Zhou et al., 2008) was proposed for estimation of the hydraulic properties of fractured rock masses at relatively lower stress level based on the superposition principle of flow dissipation energy. It was shown that the model is equivalent to Snow's model (Snow, 1969) and Oda's model (Oda, 1986) not only in form but also in function when considering the effects of mechanical loading process on the evolution of hydraulic properties. This model relies on the geometrical characteristics of rock fractures and the corresponding fracture network, and demonstrates the coupling effect between fluid flow and deformation. In this model, the pre-peak dilation and contraction effect of the fractures under shear loading is also empirically considered. It was applied to estimate the hydraulic properties of the rock mass in the dam site of the Laxiwa Hydropower Project located in the upstream of the Yellow River, China, and the model predictions have a good agreement with the site observations from a large number of singlehole packer tests.

The strain-dependent hydraulic conductivity model (Chen et al., 2007), on the other hand, was established by an equivalent non-associative elastic-perfectly plastic constitutive model with mobilized dilatancy to characterize the nonlinear mechanical behavior of fractured rock masses under complex loading conditions and to separate the deformation of weaker fractures from the overall deformation response of the homogenized rock masses. The major advantages of the model lie in the facts that the proposed hydraulic conductivity tensor is related to strains rather than stresses, hence enabling hydro-mechanical coupling analysis to include the effect of material nonlinearity and post-peak dilatancy, and the proposed model is easy to be included in a FEM code, particularly suitable for numerical analysis of hydromechanical problems in rock engineering with large scales. Numerical simulations were performed to investigate the changes in hydraulic conductivities of a cube of fractured rock mass under triaxial compression and shear loading as well as an underground circular excavation in biaxial stress field at the Stripa mine (Kelsall et al., 1984; Pusch, 1989), and the simulation results are justified by in-situ experimental observations and compared with Liu's elastic strain-dependent analytical solution (Liu et al., 1999).

Unless otherwise noted, continuum mechanics convention is adopted in this chapter, i.e., tensile stresses are positive while compressive stresses are negative. The symbol (:) denotes an inner product of two second-order tensors (e.g., $a{:}b{=}a_{ij}b_{ij}$) or a double contraction of adjacent indices of tensors of rank two and higher (e.g., $c{:}d{=}c_{ijkl}d_{kl}$), and ($\otimes$) denotes a dyadic product of two vectors (e.g., $a{\otimes}b{=}a_ib_j$) or two second-order tensors (e.g., $c{\otimes}d{=}c_{ij}d_{kl}$).

STRESS-DEPENDENT HYDRAULIC CONDUCTIVITY OF ROCK FRACTURES

In this section, the elastic deformation behavior of rock fractures at the pre-peak loading region will be first presented, and then a stress-dependent hydraulic conductivity model will be formulated. The deformation model (or indirectly the hydraulic conductivity model) is validated by the laboratory shear-flow coupling test data obtained by Liu et al. (2002). The main purpose of this section is to provide a theory for developing a stress-dependent hydraulic conductivity tensor for fractured rock masses that will be presented later in Section 4.

Characterization of Rock Fractures

One of the major factors that govern the flow behavior through fractured rocks is the void geometry, which can be described by several geometrical parameters, such as aperture, orientation, location, size, frequency distribution, spatial correlation, connectivity, and contact area, etc. (Olsson & Barton, 2001; Zhou et al., 1997; Zhou & Xiong, 1997). Real fractures are neither so solid as intact rocks nor void only. They have complex surfaces and variable apertures, but to make the flow analysis tractable, the geometrical description is usually simplified. It is common to assume that individual fractures lie in a single plane and have a constant hydraulic aperture. When the fractures are subjected to normal and shear loadings, the fracture aperture, the contact area and the matching between the two opposing surfaces will be altered. As a result, the equivalent hydraulic aperture of the fractures varies with their normal and shear stresses/displacements, which demonstrates the apparent coupling mechanism between fluid flow and stress/deformation (Min et al., 2004).

The aperture of rock fractures tends to be closed under applied normal compressive stress. The asperities of the surfaces will be crushed when their localized compressive stresses exceed their compressive strength. As a large number of asperities are crushed under high compressive stress, the contact area between the fracture walls increases remarkably and the crushed rock particles partially or fully fill the nearby void, which decreases the effective flow area, reduces the hydraulic conductivity of the fracture, and even changes the flow paths through fracture plane. Fig. 1 depicts the increase in contact area of fractures under increasing compressive stresses modelled by boundary element method (Zimmerman et al., 1991).

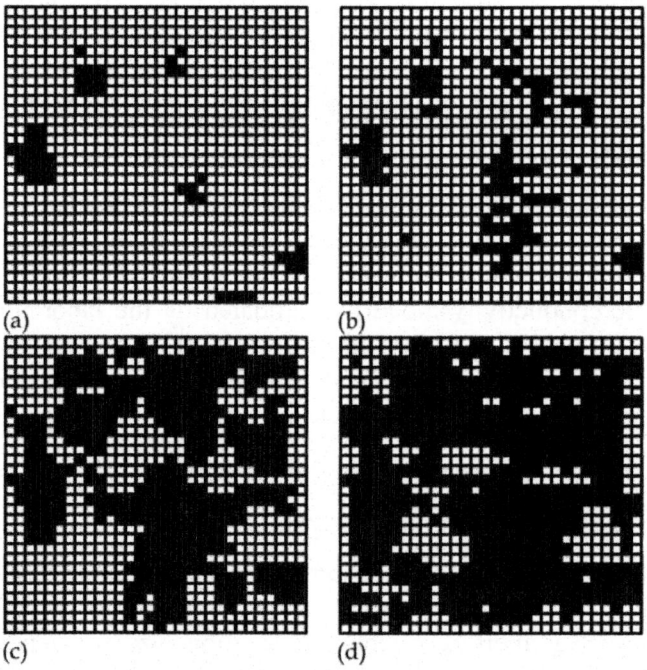

Figure 1. Variation of contact surface of fractures under increasing compressive stresses (after Zimmerman et al., (1991): (a) P=0 MPa; (b) P=20 MPa; (c) P=40 MPa and (d) P=60 MPa.

The coupling process between fluid flow and shear deformation is more related to the roughness of fractures and the matching of the constituent walls. Fig. 2 shows the impact of the fracture structure on the shear stress-deformation coupling mechanism. In Fig. 2(a), the opposing walls of the fracture are well matched so that the fracture always dilates and the hydraulic conductivity increases under shear loading as long as the applied normal stress is not high enough for the asperities to be crushed. For the state shown in Fig. 2(c), shear loading will result in the closure of the fracture and the reduction in hydraulic conductivity. Fig. 2(b) illustrates a middle state between (a) and (c), and its shearing effect depends on the direction of shear stress. When the matching of a fracture changes from (a) to (b) then to (c) under shear loading, shear dilation occurs. On the other hand, shear contraction takes place from the movement of the matching from (c) to (b) then to (a). In a more complex scenario, shear dilation and shear contraction may happen alternately, resulting in the fluctuation of the hydraulic behavior of the fractures.

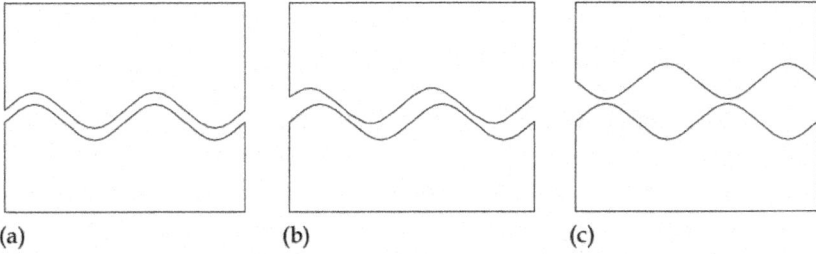

(a)   (b)   (c)

Figure 2. Shear dilation and shear contraction of fractures: (a) well-matched; (b) fair-matched; and (c) bad-matched.

An elastic constitutive model for rock fractures To formulate the stress-dependent hydraulic conductivity for rock fractures, we model the fractures by an interfacial layer, as shown in Fig. 3. The interfacial layer is a thin layer with complex constituents and textures (depending on the fillings, asperities and the contact area between its two opposing walls). Assumption is made here that the apparent mechanical response of the interfacial layer can be described by Lame's constant λ and shear modulus μ. Because the thickness of the interfacial layer (i.e., the initial mechanical aperture of the fracture) is generally rather small comparing to the size of rock matrix, it is reasonable to assume that $\varepsilon_x = \varepsilon_y = 0$ and $\gamma_{xy} = \gamma_{yx} = 0$ within the interfacial layer. Then according to the Hooke's law of elasticity, the elastic constitutive relation for the interfacial layer under normal stress σ_n and shear stress τ can be written in the following incremental form:

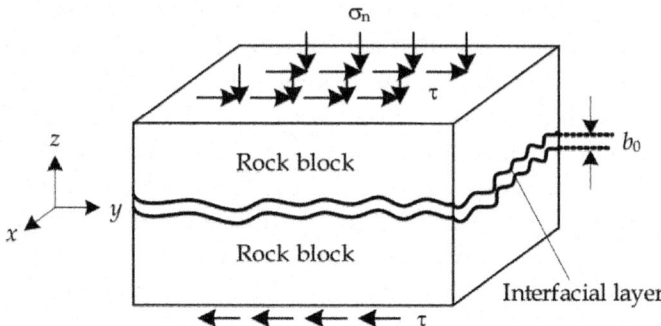

Figure 3. The interfacial layer model for rock fractures.

$$\begin{Bmatrix} d\sigma'_n \\ d\tau \end{Bmatrix} = \begin{bmatrix} \lambda + 2\mu & 0 \\ 0 & \mu \end{bmatrix} \begin{Bmatrix} d\varepsilon_n \\ d\gamma \end{Bmatrix} \tag{1}$$

For convenience, we use u_1 to denote the relative normal displacement of the interfacial layer caused by the effective normal stress σ'_n, δ to denote the relative tangential displacement caused by the shear stress τ, and u_2 to denote the relative normal displacement caused by shear dilation or contraction (positive for dilatant shear, negative for contractive shear). Hence, the total normal relative displacement u is represented as

$$u = u_1 + u_2 \tag{2}$$

The increments of strains, $d\varepsilon_n$ and $d\gamma$, can be expressed in terms of the increments of relative displacements, du_1 and $d\delta$, as follows:

$$\begin{cases} d\varepsilon_n = du_1 / (b_0 + u) \\ d\gamma = d\delta / (b_0 + u) \end{cases} \tag{3}$$

where b_0 is the thickness of the interfacial layer or the initial mechanical aperture of the fracture. Substituting Eq. (3) in Eq. (1) yields:

$$\begin{Bmatrix} d\sigma'_n \\ d\tau \end{Bmatrix} = \begin{bmatrix} k_n & 0 \\ 0 & k_s \end{bmatrix} \begin{Bmatrix} du_1 \\ d\delta \end{Bmatrix} \tag{4}$$

where k_n and k_s denote the tangential normal stiffness and tangential shear stiffness of the interfacial layer, respectively.

$$k_n = (\lambda + 2\mu) / (b_0 + u), \quad k_s = \mu / (b_0 + u) \tag{5}$$

Interestingly, k_n and k_s show a hyperbolic relation with normal deformation and characterize the deformation response of the interfacial layer under the idealized conditions that each fracture is replaced by two smooth parallel planar plates connected by two springs with stiffness values k_n and k_s. As can be seen from Eq. (5), as long as the initial normal stiffness and shear stiffness with zero normal displacement, k_{n0} and k_{s0}, are known, they can be used as substitutes for λ and μ.

Substituting Eq. (2) in Eq. (4) results in:

$$d\sigma'_n = \frac{(\lambda + 2\mu)du_1}{b_0 + u_1 + u_2} \tag{6}$$

$$d\tau = \frac{\mu d\delta}{b_0 + u_1 + u_2} \tag{7}$$

Suppose normal stress σ_n is firstly applied before the loading of shear stress, u_1 can be obtained by directly integrating Eq. (6):

$$u_1 = (b_0 + u_2)\left[\exp\left(\frac{\sigma'_n}{\lambda + 2\mu}\right) - 1\right] \tag{8}$$

Here, it is to be noted that the elastic constitutive model for the rock fracture leads to an exponential relationship between the fracture closure and the applied normal stress, which has been widely revealed in the literature, e.g., in Min et al. (2004).

On the other hand, the shear expansion caused by $d\delta$ can be estimated from shear dilation angle d_m:

$$du_2 = \tan d_m \, d\delta \tag{9}$$

By introducing two parameters, s and φ, pertinent to normal stress σ_n, we represent the dilation angle d_m under normal stress σ_n in the form of Barton's strength criterion for joints (Barton, 1976) ($\tau = \sigma_n \tan(2d_m + \varphi_b)$, where φ_b is the basic frictional angle of joints):

$$\tan d_m = \frac{1}{2}\left[\arctan\left(\frac{\tau}{s}\right) - \varphi\right] \tag{10}$$

Obviously, s is a normal stress-like parameter, and φ is a frictional angle-like parameter. But to make the above formulation still valid into pre-dilation state (i.e., shear contraction state), s and φ differ from their initial implications. Later, we will show how they can be back calculated from shear experimental data.

Substituting Eqs. (9) and (10) into (7) yields:

$$\frac{du_2}{b_0 + u_1 + u_2} = \frac{1}{2\mu}\left[arctan\left(\frac{\tau}{s}\right) - \varphi\right]d\tau$$

(11)

By integrating Eq. (11), we have:

$$u_2 = (b_0 + u_1)\left\{exp\left[\frac{|\tau|}{2\mu}\left(arctan\frac{|\tau|}{s} - \varphi\right) - \frac{s}{4\mu}\ln\left(1 + \frac{\tau^2}{s^2}\right)\right] - 1\right\}$$

(12)

By solving the simultaneous equations, Eqs. (8) and (12), we have:

$$\begin{cases} u_1 = \dfrac{A(1+B)}{1-AB}b_0 \\[2mm] u_2 = \dfrac{B(1+A)}{1-AB}b_0 \end{cases}$$

(13)

where

$$A = exp\left(\frac{\sigma'_n}{\lambda + 2\mu}\right) - 1$$

(14)

$$B = exp\left[\frac{|\tau|}{2\mu}\left(arctan\frac{|\tau|}{s} - \varphi\right) - \frac{s}{4\mu}\ln\left(1 + \frac{\tau^2}{s^2}\right)\right] - 1$$

(15)

Thus, the total normal deformation under normal and shear loading can be obtained,

$$u = u_1 + u_2 = \frac{A + B + 2AB}{1 - AB}b_0$$

(16)

The actual aperture of the fracture, $b = b_0 + u$, is given by:

$$b = b_0 + u = (1 + \chi)b_0$$

(17)

where

$$\chi = \frac{A + B + 2AB}{1 - AB}$$

(18)

Stress-Dependent Hydraulic Conductivity for Rock Fractures

Since natural fractures have rough walls and asperity areas, it is not appropriate to directly use the aperture derived by Eq. (17) for describing the hydraulic conductivity of the fractures. Instead, an equivalent hydraulic aperture is usually taken to represent the percolation property of the fractures, as demonstrated in Section 1. Based on experimental data, the relationship between the equivalent hydraulic aperture and the mechanical aperture has been widely examined in the literature, and the empirical relations proposed by Lomize (1951), Louis (1971), Patir & Cheng (1978), Barton el al. (1985) and Olsson & Barton (2001) are listed in Table 1. For example, if Patir and Cheng's model is used to estimate the equivalent hydraulic aperture that accounts for the flow-deformation coupling effect in pre-peak shearing stage, then there is

$$b^* = (1+\chi)b_0[1-0.9\exp(-0.56/C_v)]^{1/3} \tag{19}$$

where C_v is the variation coefficient of the mechanical aperture of the discontinuities, which is mathematically defined as the ratio of the root mean squared deviation to the arithmetic mean of the aperture. For convenience, Eq. (19) is rewritten as:

$$b^* = b_0 f(\beta) \tag{20}$$

Obviously, f(β) is a function of the normal and shear loadings, the mechanical characteristics and the aperture statistics of the fractures.

Thus, the hydraulic conductivity of the fractures subjected to normal and shear loadings is approximated by the hydraulic conductivity of the laminar flow through a pair of smooth parallel plates with infinite dimensions:

$$k = \frac{gb^{*2}}{12v} \tag{21}$$

where k is the hydraulic conductivity, g is the gravitational acceleration, and v is the kinematic viscosity of the fluid.

An alternative approach to account for the deviation of the real fractures from the ideal conditions assumed in the parallel smooth plate theory is to adopt a dimensionless constant, ς, to replace the constant multiplier, 1/12, in Eq. (21), where $0<\varsigma<1/12$ (Oda, 1986). In this manner, the hydraulic conductivity of the fractures is estimated by

$$k = \varsigma \frac{gb^2}{v} \tag{22}$$

Clearly, the constant, ς, approaches 1/12 with increasing scale and decreasing roughness of the fractures.

Eqs. (21) and (22) show that the hydraulic conductivity of a rock fracture varies quadratically with its mechanical aperture. The latter depends, by Eq. (18), on the normal and shear stresses applied on the fracture. Hence, we call the established model, Eq. (21) or (22), the stress-dependent hydraulic conductivity model, and it is suitable to describe the hydraulic behavior of the fractures subjected to mechanical loading in the pre-peak stage.

Table 1. Empirical relations between equivalent hydraulic aperture and mechanical aperture

Authors	Expressions	Descriptions
Lomize (1951)	$b^* = b\left[1.0 + 6.0(e/b)^{1.5}\right]^{-1/3}$	b^* is the equivalent hydraulic aperture of fractures, b the mechanical aperture, e the absolute
Louis (1971)	$b^* = b\left[1.0 + 8.8(e_m/D_H)^{1.5}\right]^{-1/3}$	asperity height, e_m the average
Patir & Cheng (1978)	$b^* = b\left[1 - 0.9\exp(-0.56/C_v)\right]^{1/3}$	asperity height, D_H the hydraulic radius, C_v the variation coefficient
Barton, et al. (1985)	$b^* = b^2 JRC^{-2.5}$	of the mechanical aperture, JRC the joint roughness coefficient, JRC_0
Olsson & Barton (2001)	$\begin{cases} b^* = b^2 JRC_0^{-2.5} & \delta \le 0.75\delta_p \\ b^* = b^{1/2} JRC_{mob} & \delta \ge \delta_p \end{cases}$	the initial value of JRC, JRC_{mob} the mobilized JRC, δ the shear displacement and δ_p the peak shear displacement.

Validation of the Elastic Constitutive Model

The key point of the stress-dependent hydraulic conductivity model is whether the established elastic constitutive model can properly describe the variation of mechanical aperture under normal and shear loadings at low stress level. Here, we use the results of the laboratory test performed by Liu et al. (2002) to validate the mechanical model. The test was conducted to study shear-flow coupling properties for a marble fracture with fillings of sand under low normal stresses and small shear displacements.

The marble specimen for shear-flow coupling test is illustrated in Fig. 4, which was collected from the Daye Iron Mine in China. The uniaxial compressive strength and density of the rock sample are 52.4 MPa and 2.66×10^3 kg/m^3, respectively. The specimen was cut into round shape and the fracture surfaces were polished, with its size of 290 mm in diameter and 200 mm in height. The opposite walls of the fracture were cemented with a layer of filtered sands with their diameters ranged from 0.5 to 0.69 mm, and the fracture was further filled with the same sands. The initial aperture of the fracture, b_0, is about 1.31 mm.

The coupled shear-flow test were conducted by first applying a prescribed normal stress ranging between 0.1 and 0.5 MPa and then applying shear displacement in steps until a maximum displacement of about 0.4 mm was reached. During tests, steady-state fluid flow rate and normal displacement were continuously recorded.

With such low normal stresses and small shear displacements, it is reasonable to consider that the fracture behaves elastic during the coupled shear-flow test. According to the experimental results, the elastic parameters, λ and μ, of the fracture with fillings are estimated as λ=1.81 MPa and μ=3.62 MPa. In order to enable Eq. (16) to predict the mechanical aperture of the facture under normal and shear loads, the normal stress-like parameter, s, and the frictional angle-like parameter, φ, should be further determined. Fortunately, both of them can be derived by fitting the experimental curve between normal displacement and shear displacement, as plotted in Fig. 5, using Eq. (16) such that the least square error is minimized. With this approach, we obtain that for σ_n=0.1 MPa, s=0.062, φ=1.324, and for σ_n=0.4 MPa, s=0.046, φ=1.310.

Fig. 5 plots the experimental results as well as the model predictions of the relation between mechanical aperture and shear displacement of the fracture under constant normal stresses. Generally, the proposed elastic constitutive model manifests the behavior of the fracture with fillings during the shear-flow coupling test with low normal and shear loads. Shear contraction is observed in the initial 0.06-0.08 mm of shear displacement, which is followed by shear dilation in the remaining of the shear displacement. This property, which is actually ensured by the empirical relation assumed in Eq. (9), suggests that the resultant model is suitable for phenomenologically describing the pre-peak shear-flow coupling effect of fractures.

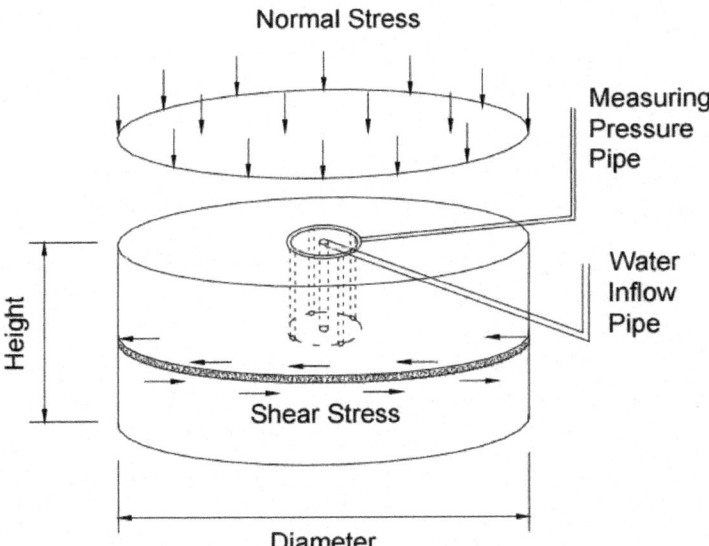

Figure 4. Sketch of the marble specimen for shear-flow coupling test.

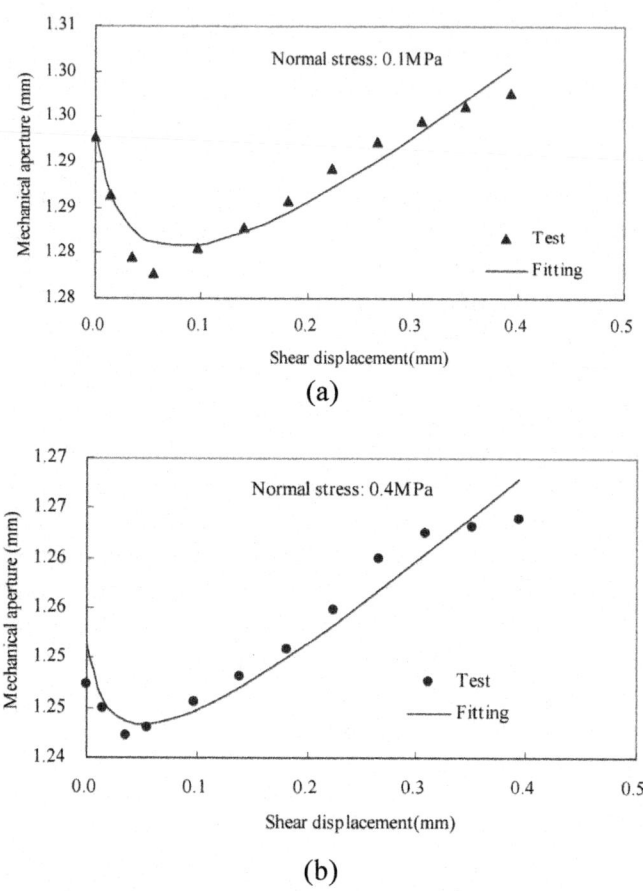

Figure 5. Mechanical aperture versus shear displacement curve under constant normal stress: (a) Normal stress: 0.1 MPa and (b) Normal stress: 0.4 MPa.

Fig. 6 further depicts the sensitivity of s and φ on the behavior of the fracture. In Fig. 6(a), φ is fixed to 1.324, while s varies from 0.02 to 0.08. As s increases, shear contraction more apparently manifests, and the mechanical aperture versus shear displacement curves become lower as a whole. On the other hand, the effect of varying φ from 0.524 to 1.222 but fixing s to 0.062 is plotted in Fig. 6(b). For small value of φ, shear contraction is trivial and the curve extends with a larger slope. As φ increases, however, shear contraction becomes relatively remarkable and the curve turns relatively flat. Thus, by adjusting s and φ, the mechanical and hydraulic behaviors of the fracture can be appropriately established.

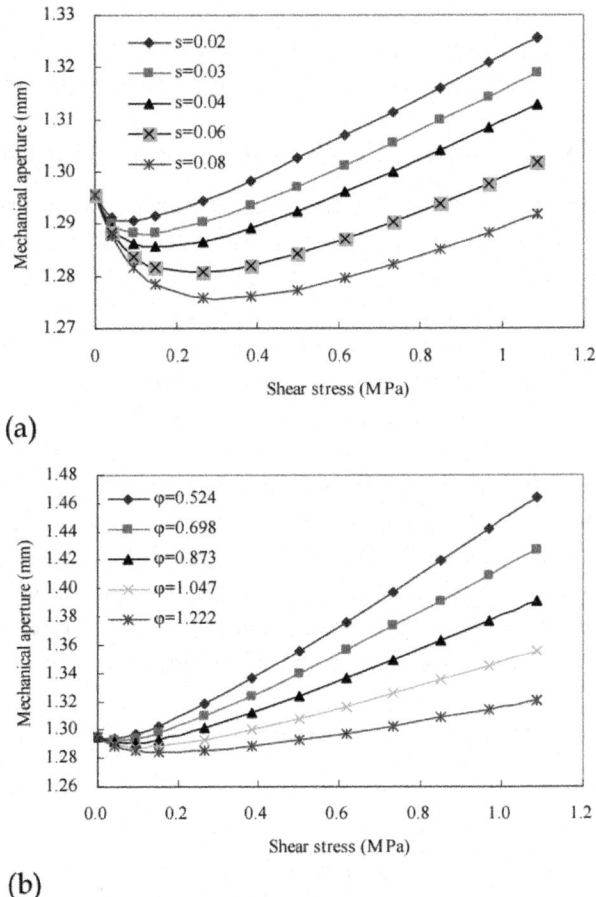

(a)

(b)

Figure 6. The sensitivity of s and φ on the behavior of the fracture: (a) φ=1.324 and (b) s=0.062.

STRAIN-DEPENDENT HYDRAULIC CONDUCTIVITY OF ROCK FRACTURES

In this section, we develop an elasto-plastic constitutive model for single hard rock fractures with consideration of nonlinear normal deformation and post-peak shear dilatancy, and then formulate the strain-dependent hydraulic conductivity for the fractures under dilated shear loading. Compared with the stress-dependent model presented in Section 2, one major difference is that the strain-dependent model is capable of describing the influence of postpeak mechanical response on the hydraulic properties of the fractures. This work is of paramount importance in the sense that the theoretical results are directly comparable with the experimental data of coupled shear-flow test, e.g. in Esaki et al. (1999). The straindependent hydraulic conductivity tensor can then be developed on this basis, which will be presented later in Section 5.

An Elasto-Plastic Constitutive Model for Rock Fractures

The underlying physical model considered is the same with the model plotted in Fig. 3, in which a fracture of hard rock is located at the mid-height of a specimen between two intact rock blocks. The height of the specimen is denoted by s, and the initial aperture of the fracture is b_0. When constant normal stress σ_n and increasing shear displacement δ are applied on the specimen, typical and idealized curves of shear displacement versus shear stress and shear displacement versus normal displacement (i.e. $\delta\sim\tau$ curve and $\delta\sim u$ curve) are plotted in Fig. 7. The shear stress increases linearly with the shear displacement (linked by the initial shear stiffness of the fracture, k_{s0}) until the shear stress approaches the peak, τ_p, which is then followed by a shear softening process in which the shear stress decreases to a residual level at a decreasing gradient with increasing shear displacement. For the purpose of deriving the hydraulic property of the fracture in post-peak loading section, however, an elastic-perfectly plastic $\delta\sim\tau$ relationship can be assumed, as shown in Fig. 7(a)

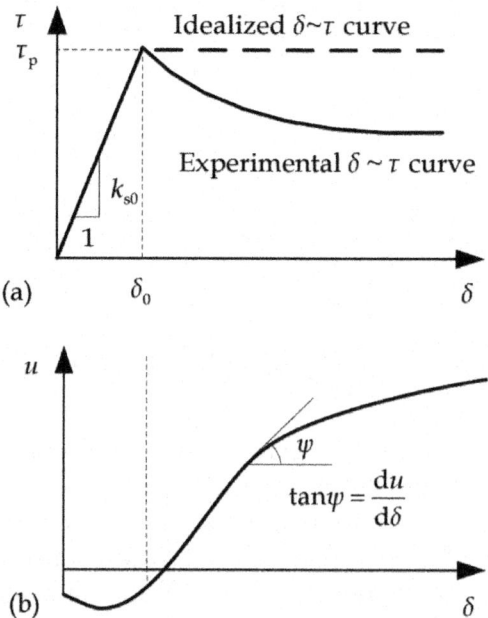

Figure 7. Typical and idealized curves of shear displacement versus shear stress and shear displacement versus normal displacement of a fracture subjected to normal and shear loads.

The deformation response of a rock fracture subjected to normal and shear loadings includes two components: one is the nonlinear closure of the fracture due to normal compression, and the other is the opening of the fracture due to shear dilation. Experimental results in Esaki et al. (1999) show that in the shearing process under constant normal loading, dilatancy will start when the shear stress approaches the peak and it increases at a decreasing gradient with increasing shear displacement, as illustrated in Fig. 7(b). As a result, the aperture of the fracture and then the hydraulic conductivity vary with increasing shear displacement.

Therefore, we may consider that shear dilatancy as well as the change in hydraulic conductivity accompanies normal and plastic shear deformations of the fracture. To deduce the hydraulic conductivity of the fracture with an averaging method, which will be further used later for deriving the hydraulic conductivity tensor for fractured rocks, we view the specimen with fracture as an equivalent continuous medium, i.e. the hydromechanical properties of the fracture are averaged into the whole specimen. As can be seen later, such a treatment does not affect our final solution to a single fracture, but it renders valid the small strain assumption on the fractures in the presence of post-sliding plasticity.

For a one-dimensional problem with a single rock fracture, the elasto-plastic constitutive model can be represented in the following forms:

$$\gamma_p = \gamma - \gamma_e = \frac{\delta}{s} - \frac{\delta_0}{s} = \frac{\delta}{s} - \frac{\tau_p}{sk_{s0}} \tag{23}$$

$$\varepsilon_n = \frac{\sigma_n'}{sk_n} + \int \tan\psi \, d\gamma_p \tag{24}$$

where γ, γ_e and γ_p are the total shear strain, the elastic shear strain and the plastic shear strain of the fracture, respectively; ε_n is the normal strain of the fracture; τ_p is the peak shear stress of the fracture under effective normal stress σ_n'; k_n and k_{s0} are, respectively, the normal stiffness and the initial shear stiffness of the fracture; δ_0 is the maximum elastic shear displacement upon shear yielding, with $\delta_0 = \tau_p/k_{s0}$, as shown in Fig. 7(a); and ψ is the mobilized dilatancy angle of the fracture. Note that in Eq. (24), the first term on the right hand side denotes the nonlinear closure of the fracture subjected to effective normal stress σ_n', while the second term denotes the opening of the fracture due to shear dilatancy.

Existing studies have indicated that shear dilatancy is highly dependent on the plasticity already experienced by the fractures and normal stress, and non-negligibly dependent on scale (Barton & Bandis, 1982; Yuan & Harrison, 2004; Alejano & Alonso, 2005). The decaying process of the dilatancy angle in line with plasticity can be described by the following negative exponential expression through the plastic shear strain, γ_p, or indirectly through the plastic shear displacement, δ, on the basis of Eq. (23):

$$\psi = \psi_{peak}\exp\left[-r(\delta - \delta_0)\right] \qquad (25)$$

where r is a parameter for modelling the rate of decay that ψ undergoes as the plastic shear strain evolves. If r=0, then a constant dilatancy angle is recovered. As r→∞, the dilatancy angle quickly decays to zero. ψ_{peak} is the peak dilatancy angle of the fracture in the form of (Barton & Bandis, 1982)

$$\psi_{peak} = JRC \cdot \log_{10}\frac{JCS}{-\sigma'_n} \qquad (26)$$

where JRC and JCS are the roughness coefficient and the wall compressive strength of fractures, respectively, and the actual values of them should be scale-corrected (Barton & Bandis, 1982). Thus, the dependencies of fracture dilatancy on plasticity, normal stress and scale are established through Eqs. (25) and (26).

Note that Eq. (25) shares the same shape with the asperity angle proposed for the description of shear dilatancy and surface degradation (Plesha, 1987), but the latter is represented as a function of the plastic tangential work. With the assumption of elasticperfectly plasticity, they are fully equivalent for monotonic loading (Jing et al., 1993). Cyclic loading is not a concern in this simple model, but when cyclic loading is involved, another independent function can be associated to the reverse loading that starts from the original point, just as the suggestion given in Plesha (1987) for asperity angles in two opposite directions, in order to satisfy the thermodynamic restriction condition presented in Jing et al. (1993).

Using the Mohr-Coulomb criteria, the peak shear stress τ_p of the fracture under effective normal stress σ'_n satisfies

$$\tau_p = -\sigma'_n\tan\varphi + c \qquad (27)$$

where φ and c are the frictional angle and the cohesion of the fracture.

Differentiating Eq. (23) yields

$$d\gamma_p = d\gamma = \frac{1}{s}d\delta \qquad (28)$$

Combining Eqs. (24) and (28) results in

$$\Delta b \approx s\varepsilon_n = \frac{\sigma'_n}{k_n} + \int_{\delta_0}^{\delta} \tan\psi(\delta)d\delta \tag{29}$$

An interesting phenomenon in Eq. (29) is, as described before, the change in the aperture of the fracture, Δb, is irrelevant to the height of the specimen, s. To conveniently use this formulation, two remedies can be further made:

First, suppose that the hyperbolic variation of k_n with the increase of aperture can be considered in the following (Huang et al., 2002):

$$k_n = \frac{-\sigma'_n + b_0 k_{n0}}{b_0} \tag{30}$$

where k_{n0} is the initial normal stiffness of the fracture.

Second, by employing the Taylor series expansion (truncated at the third order term), $\tan\psi$ can be adequately approximated by $\psi + \psi^3/3$ in radians for a rather large ψ_{peak}, e.g. 30°.

From Eq. (29) and the above two remedies, we have

$$\Delta b = \chi b_0 \tag{31}$$

$$b = b_0 + \Delta b = (1+\chi)b_0 \tag{32}$$

with the parameter, χ, in the following form

$$\chi = \frac{\sigma'_n}{-\sigma'_n + b_0 k_{n0}} + \frac{1}{b_0}\left\{ \frac{\psi_{peak}}{r}\left[1 - e^{-r(\delta-\delta_0)}\right] + \frac{\psi_{peak}^3}{9r}\left[1 - e^{-3r(\delta-\delta_0)}\right] \right\} \tag{33}$$

Strain-Dependent Hydraulic Conductivity for Rock Fractures

Rewrite from Eq. (22) the initial hydraulic conductivity of the fracture, k_0, in the following form:

$$k_0 = \varsigma\frac{g b_0^2}{\nu} \tag{34}$$

Then, the hydraulic conductivity of the fracture under effective normal stress σ'_n and shear displacement δ can be described by

$$k = \varsigma\frac{g b^2}{\nu} = k_0(1+\chi)^2 \tag{35}$$

Hence, a theoretical model of the hydraulic conductivity for a single rock fracture is finally formulated, which is totally determined by the effective normal stress σ'_n and the shear displacement δ, as well as a set of parameters characterizing the behavior of the fracture (i.e. b_0, ς, k_{n0}, k_{s0}, φ, c, JRC, JCS and r, which all can be deduced or back-calculated from experimental data).

Note that by Eqs. (35) and (33), the proposed hydraulic conductivity model for rock fractures subjected to normal and shear loadings with mobilized dilatancy behavior depends in form on the plastic shear displacement, but from Eq. (23), one observes that the model depends indirectly on the plastic shear strain. Thus, we classify the established model into the stain-dependent hydraulic conductivity model.

Validation of the Proposed Model

Esaki et al. (1999) systematically investigated the coupled effect of shear deformation and dilatancy on hydraulic conductivity of rock fractures by developing a new laboratory technique for coupled shear-flow tests of rock fractures. In this section, we validate the theory proposed in Section 3.2 using the experimental data reported in Esaki et al. (1999). For this purpose, we first briefly introduce the experiments, and then predict our analytical results through Eqs. (31) and (35) by directly comparing with the experimental data.

The Coupled Shear-Flow Tests

The coupled shear-flow tests were conducted with an artificially created granite fracture sample under various constant normal loads and up to a residual shear displacement of 20 mm (Esaki et al., 1999). The underlying specimen for coupled shear-flow tests is sketched in Fig. 3, with its size of 120 mm in length, 100 mm in width and 80 mm in height. The initial aperture of the created fracture, b_0, is about 0.15 mm. The value of JRC is 9, and the value of JCS is 162 MPa, respectively.

The coupled shear-flow tests were conducted by first applying a prescribed normal stress ranging between 1 MPa and 20 MPa and then applying shear displacement in steps at a rate of 0.1 mm/s until a maximum shear displacement of 20 mm was reached. During tests, steady-state fluid flow rate, shear loading and dilatancy were all continuously recorded. The hydraulic aperture and conductivity were back-calculated by applying the cubic law, with the flow equations solved by using a finite difference method.

Determination of the Parameters for the Proposed Model

Some of the experimental values of the mechanical parameters of the fracture specimen during the coupled shear-flow tests are listed in Table 2 (taken from Table 1 in Esaki et al. (1999)). Using the data as listed in Table 2, we plot the peak shear stress versus normal stress curve in Fig. 8, which can be fitted by a linear equation $\tau_p = 1.058\sigma_n + 0.993$ with a high correlation coefficient of 0.9999. Therefore, the shear strength of the specimen can be derived as $\varphi = 46.6°$ and $c = 0.99$ MPa, respectively.

Table 2. Mechanical parameters of the artificial fracture (After Esaki et al. (1999))

σ_n (MPa)	τ_p (MPa)	k_{s0} (MPa/mm)
1	2.06	3.37
5	6.16	10.65
10	11.74	11.97
20	22.10	17.97

The initial normal stiffness of the fracture of the specimen, k_{n0}, has to be estimated from the recorded initial normal displacement with zero shear displacement under different normal stresses. From the data plotted in Fig. 9 (which is taken from Fig. 7b in Esaki et al. (1999)), k_{n0} can be estimated as $k_{n0} = 100$ MPa/mm by considering the possible deformation of the intact rock under high normal stresses. It is to be noted that in the remainder of this section, the hard intact rock deformation of the small specimen is neglected, meaning that the normal displacement of the specimen mainly occurs in the fracture of the specimen and it is approximately equal to the increment of the mechanical aperture of the fracture.

Theoretically, the decay coefficient of the fracture dilatancy angle, r, can be directly measured from the normal displacement versus shear displacement curves as plotted in Fig. 9. A better alternative, however, is to fit the experimental curves using Eq. (31) such that the least square error is minimized. By this approach, we obtain that r=0.13 with a correlation coefficient of 0.9538.

To obtain the dimensionless constant, ς, in Eq. (35) that relates the mechanical aperture to the hydraulic conductivity of the fracture under testing, further efforts are needed. A simple approach is to back-calculate ς directly using Eq. (34) with initial hydraulic conductivity, k0. But similarly, the better alternative is to fit the hydraulic conductivity versus shear displacement curves, as plotted in Fig. 11 (which is taken from Fig. 7c-f in Esaki et al. (1999)), using Eq. (35) such that the least square error is minimized. With such a method, we obtain that $\varsigma = 0.00875$. This means that the mechanical aperture, b, and the hydraulic aperture, b^*, are linked with $b^* = 0.324b$, which is very close to the experimental result shown in Fig. 8 in Esaki et al. (1999).

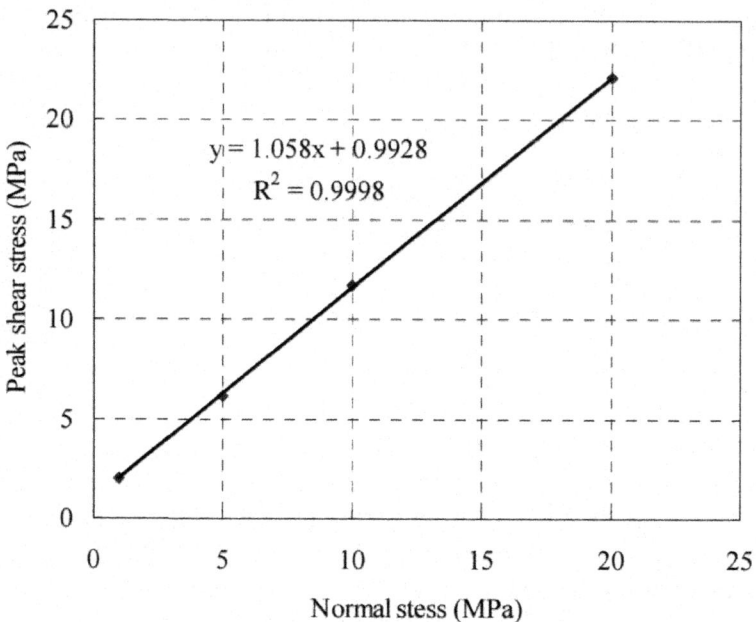

Figure 8. Peak shear stress versus normal stress curve of the fracture.

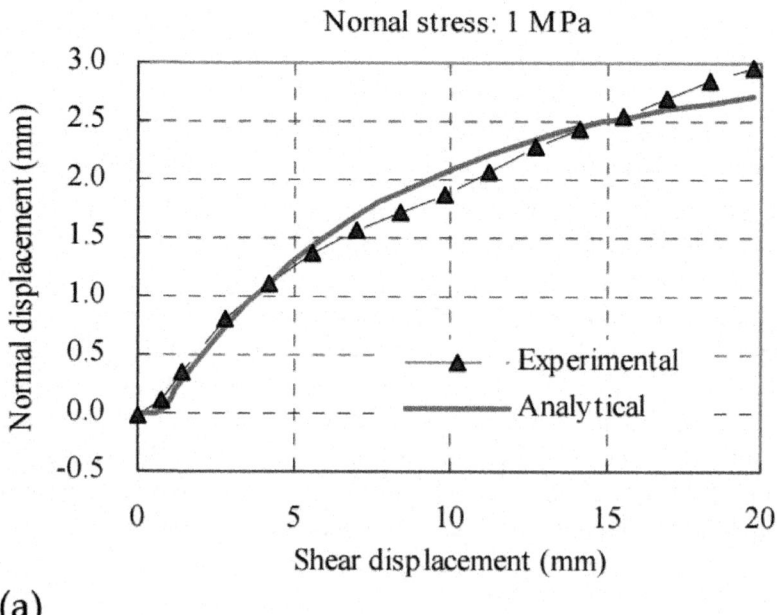

(a)

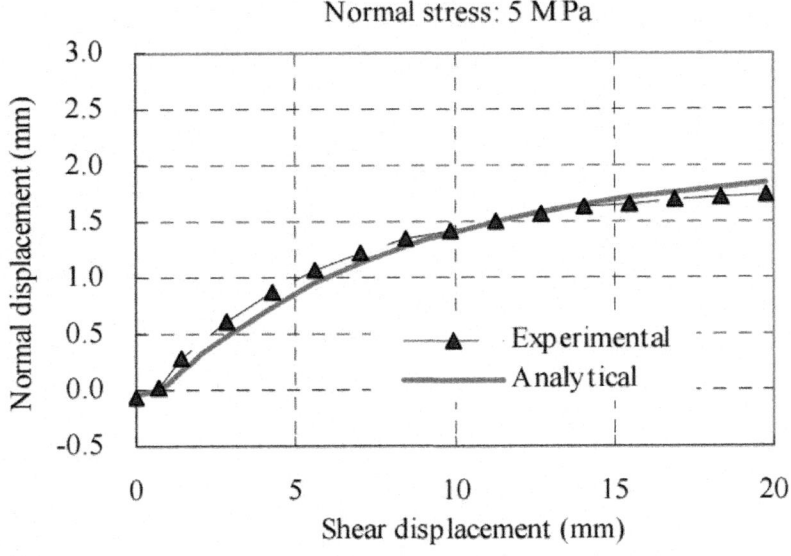

(b)

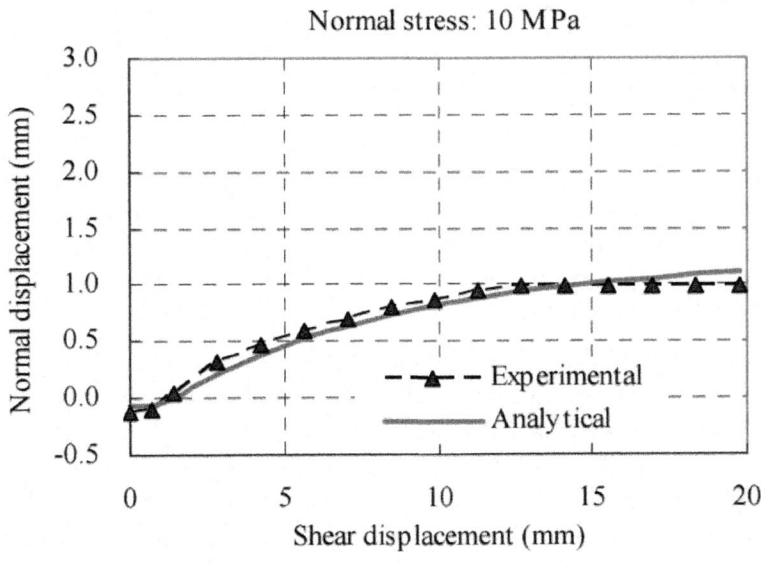

(c)

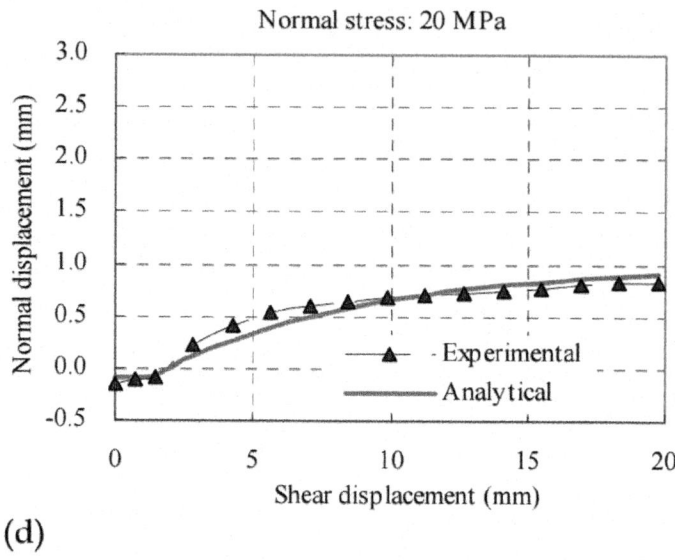

(d)

Figure 9. Comparison of the fracture aperture analytically predicted by Eq. (31) with that measured in coupled shear-flow tests.

Validation of the Proposed Theory

With the necessary parameters obtained in Section 3.3.2, we are now ready to compare the proposed model in Eqs. (31) and (35) with the experimental data presented in Esaki et al. (1999). Note that although the experimental data are available for one cycle of forward and reverse shearing, only the results for the forward shearing part are considered. The reverse shearing process, however, can be similarly modelled.

Fig. 9 depicts the relations between the mechanical aperture and shear displacement that were measured from the coupled shear-flow tests presented in Esaki et al. (1999) and predicted by using the proposed model given in Eq. (31) under different normal stresses applied during the testing. It can be observed from Fig. 9 that our proposed analytical model is able to describe the shear dilatancy behavior of a real fracture under wide range of normal stresses between 1 MPa and 20 MPa by feeding appropriate parameters. Even the fracture aperture increases by one order of magnitude due to shear dilation, the analytical model still fitted the experimental results well. For practical uses, the slight discrepancies between the analytical results and the experimental data are negligible and the proposed model is accurate enough to characterize the significant dilatancy behavior of a real fracture. This performance is largely attributed to the dilatancy model introduced through Eqs. (25) and (26). The dilatancy angles of the fracture evolving with the plastic shear displacement under different normal stresses are illustrated in Fig. 10. The high dependencies of the dilatancy angle of the fracture on normal stress and plasticity are clearly demonstrated in the curves. The peak dilatancy angle, which can be rather accurately modelled by Barton's peak dilatancy relation (Barton & Bandis, 1982), decreases logarithmically with the increase of the applied normal stress. For normal stresses of 1 MPa, 5 MPa, 10 MPa and 20 MPa, the peak dilatancy angles are 19.9°, 13.6°, 10.9° and 8.2°, respectively. On the other hand, the dilatancy angle undergoes negative exponential decay with increasing plastic shear displacement, a process related to surface degradation of rough fractures.

Fig. 11 shows the hydraulic conductivity versus shear displacement relations that were back-calculated from fluid flow results using the finite difference method from the coupled shear-flow tests presented in Esaki et al. (1999) and that are predicted by the proposed model given in Eq. (35) under different normal stresses during testing. As shown in the semi-logarithmic graphs in Fig. 11, the proposed analytical model can well predict the evolution of hydraulic conductivity of the tested rock fracture, with the change in the magnitude of 2 orders, during coupled shear-flow tests under different normal stresses. The ratios of the predicted hydraulic conductivities to the corresponding experimental results all fall in between 0.3 and 3.0, indicating that they are rather close in orders of magnitude and the predicted results are suitable for practical use.

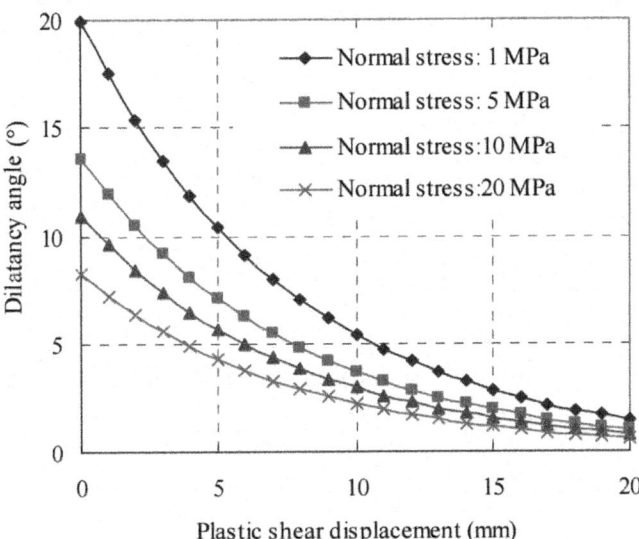

Figure 10. Dilatancy angles of the fracture evolving with the plastic shear displacement under different normal stresses.

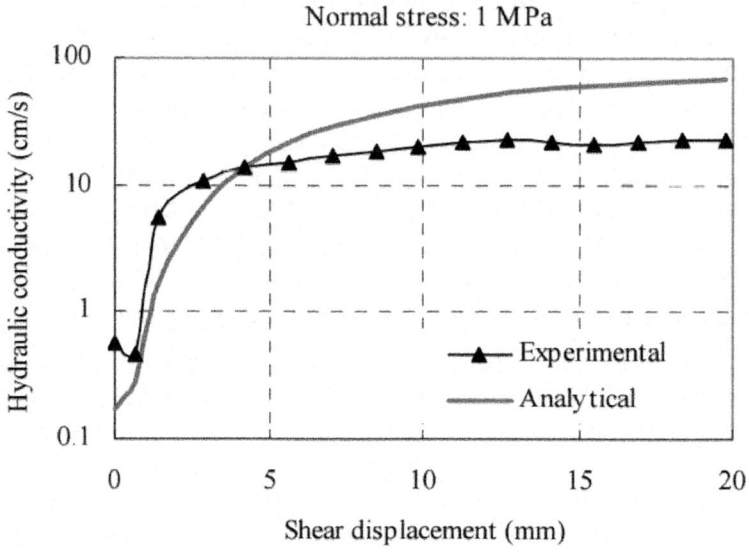

(a)

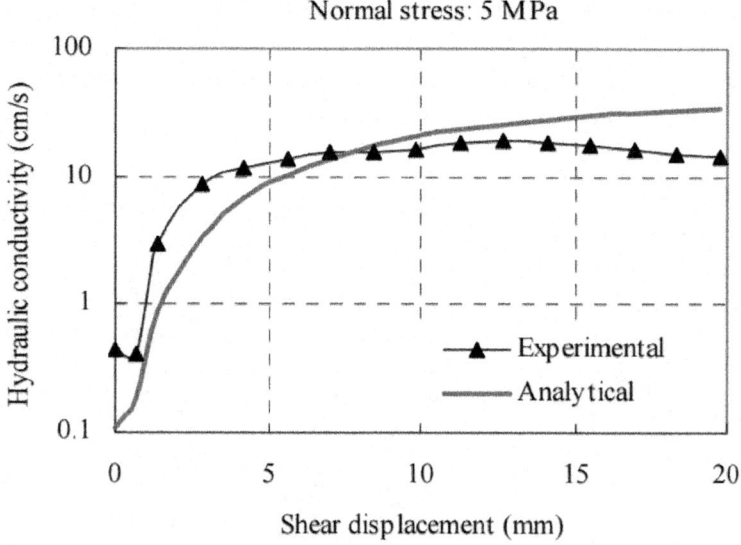

(b)

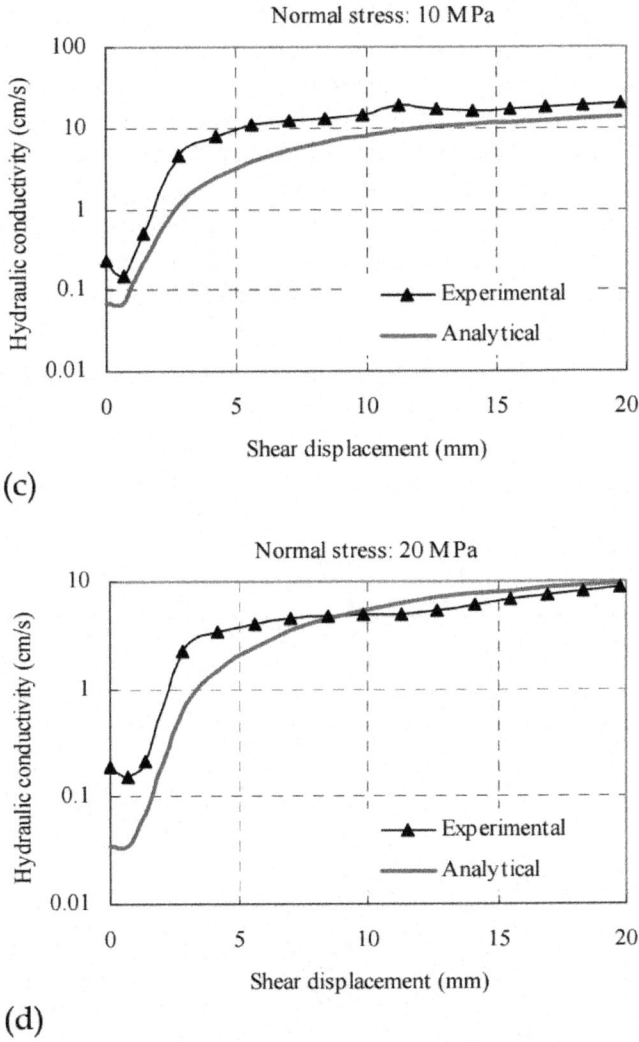

Figure 11. Comparison of the hydraulic conductivity analytically predicted by Eq. (35) with that calculated from coupled shear-flow tests with finite difference method.

STRESS-DEPENDENT HYDRAULIC CONDUCTIVITY TENSOR OF FRACTURED ROCKS

When the response of each fracture under normal and shear loading is understood (see Section 2), the remaining problem is how to formulate the hydraulic conductivity for fractured rock mass based on the geometry of the underlying fracture network. Fig. 12 depicts a two-dimensional fracture network (taken after Min et al. (2004)) in a biaxial stress field. As shown in Fig. 12, each fracture plays a role in the hydraulic conductivity of the rock mass, and its contribution primarily depends on its stress state, its occurrence, as well as its connectivity with other fractures. Also shown in Fig. 12 is the scale effect of the rock mass on hydraulic properties. When the size of the rock mass is small, only a few number of fractures are included and heterogeneity of the hydraulic conductivity of the rock mass may dominate. As the population of factures grows with the increasing size, an upscaling scheme may be available to derive a representative hydraulic conductivity tensor for the rock mass at the macroscopic scale.

Based on the above observations, in this section, we formulate an equivalent hydraulic conductivity tensor for fractured rock mass based on the superposition principle of liquid dissipation energy, in which the concept of REV is integrated and the applicability of an equivalent continuum approach is able to be validated.

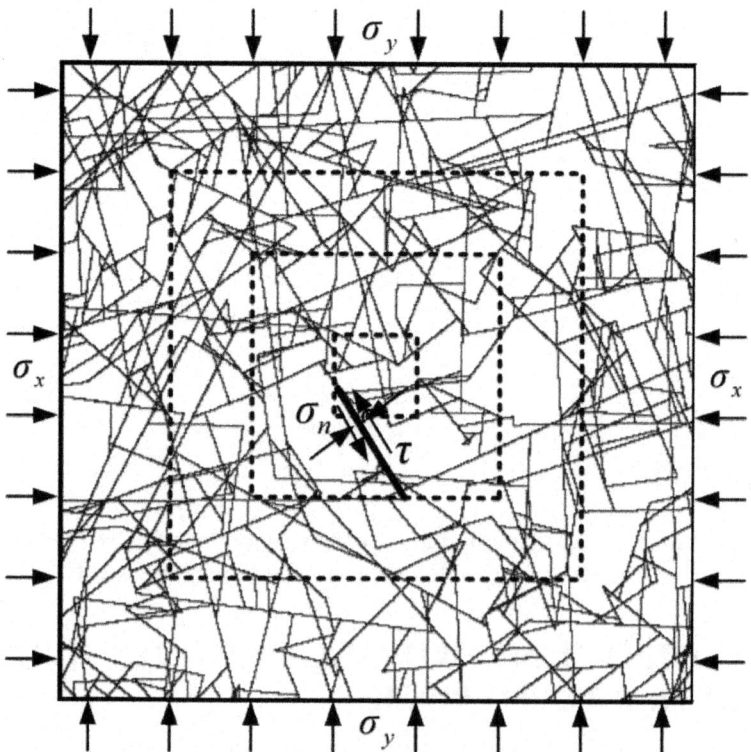

Figure 12. A fracture network (taken after Min et al. (2004)) in biaxial stress field and the scale effect of the rock mass.

Computational Model

Without loss of generality, the global coordinate system $X_1 X_2 X_3$ is established in such a way that its X_1-axis points towards the East, X_2-axis toward the North and X_3-axis vertically upward. A local coordinate system $x_1^f x_2^f x_3^f$ is associated with the *f*th set of fractures such that the x_1^f-axis is along the main dip direction, the x_2^f-axis is in the strike, and the x_3^f-axis is normal to the fractures, as shown in Fig. 13.

In order to formulate the stress-dependent hydraulic conductivity tensor for fractured rock masses using the aforementioned elastic constitutive model for rock fractures, the following assumptions, similar to Oda (1986), are made in this section:

- A cube of volume, V_p, is considered as the flow region of interest, which is cut by n sets of fractures. The orientation of each set of fractures is indicated by a mean azimuth angle β and a mean dip angleα. Other geometrical statistics of the fractures are assumed to be available through field measurements or empirical estimations.
- Even though the geometry of real fractures is complex, generally it can be simplified as a thin interfacial layer with radius r and aperture b*.
- The rock mass is regarded as an equivalent continuum medium, which means the representative elementary volume (REV) exists in the rock mass and its size is smaller than or equal to V_p.

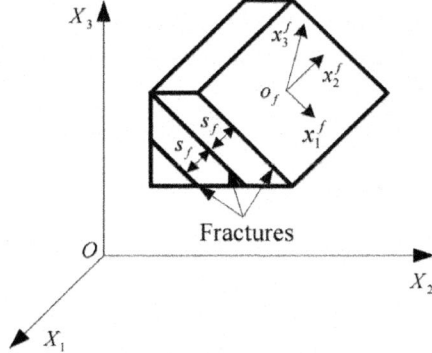

Figure 13. Coordinate systems.

Stress-Dependent Hydraulic Conductivity Tensor

Fluid flow through the equivalent continuum media can be described by the generalized 3- D Darcy's law as follows:

$$v = \mathbf{KJ} \tag{36}$$

where v denotes the vector of flow velocities, J denotes the vector of hydraulic gradients, and K is the hydraulic conductivity tensor for the rock mass.

For steady-state seepage flow, the dissipation energy density, $e(X_1, X_2, X_3)$, of fluid flow through the media can be represented as (Indelman & Dagan, 1993):

$$e = \frac{1}{2}\mathbf{J}^{\mathrm{T}}\mathbf{KJ} \tag{37}$$

Hence, the total flow dissipation energy, E, in the rock mass V_p can be calculated by performing an integration throughout the whole flow domain:

$$E = \int_{V_p} e \, d\Omega = \frac{1}{2}\int_{V_p} \mathbf{J}^{\mathrm{T}}\mathbf{KJ} \, d\Omega \tag{38}$$

If REV does exist in the rock mass and its size is smaller than or equal to V_p, by defining $\overline{\mathbf{J}}$ to be the vector of the average hydraulic gradient within V_p and $\overline{\mathbf{K}}$ to be the average hydraulic conductivity tensor, Eq. (38) can be reduced to:

$$E = \frac{1}{2}\overline{\mathbf{J}}^{\mathrm{T}}\overline{\mathbf{K}}\overline{\mathbf{J}}V_p \tag{39}$$

Suppose that the volume density of the ith set of fractures is J_{vi}. The number of this set of fractures can be estimated by $m_i = J_{vi} V_p$.

For permeable rock matrix, the flow dissipation energy shown in Eq. (39) consists of two components, i.e., the flow dissipation energy through rock matrix, E_r, and the flow dissipation energy through crack network, E_c:

$$E = E_r + E_c \tag{40}$$

E_r can be represented as:

$$E_r = \frac{1}{2}\overline{\mathbf{J}}^{\mathrm{T}}\overline{\mathbf{K}}_r\overline{\mathbf{J}}V_p \tag{41}$$

where $\overline{\mathbf{K}}_r$ denotes the hydraulic conductivity tensor for rock matrix. If rock matrix is impermeable, all elements in $\overline{\mathbf{K}}_r$ vanish. To estimate E_c, we introduce a weight coefficient W_{ij} to describe the effect of the connectivity of the fracture network on fluid flow:

$$W_{ij} = \xi_{ij} / \overline{\xi}_i \tag{42}$$

where ξ_{ij} is a stochastic variable denoting the number of fractures intersected by the jth fracture belonging to the ith set; and $\bar{\xi}_i$ denotes the maximum number of fractures cut by the ith set of fractures. Obviously, $0 \le W_{ij} \le 1$ and when $\xi_{ij} = 0$, $W_{ij} = 0$. This implies that an entirely isolated fracture which does not intersect any other fracture effectively contributes nothing to the hydraulic conductivity of the total rock mass.

For the jth fracture belonging to the ith set, a void volume equal to $\pi r_{ij}^2 b_{ij}^*$ is associated with it. Then, the flow dissipation energy through it is described as:

$$E_{cij} = W_{ij} e_{ij} \pi r_{ij}^2 b_{ij}^*$$

(43)

where e_{ij} is shown as follows:

$$e_{ij} = \frac{1}{2} k_{ij} \overline{J}_{ci}^{T} \overline{J}_{ci}$$

(44)

where k_{ij} denotes the hydraulic conductivity of the jth fracture of the ith set, which can be calculated by the stress-dependent hydraulic conductivity model, Eq. (21).

$\overline{J}_{ci}$ denotes the hydraulic gradient within the ith set of fractures:

$$\overline{J}_{ci} = (\delta - n_i \otimes n_i)\overline{J}$$

(45)

where δ is the Kronecker delta tensor, and n_i denotes the unit vector normal to the ith set of fractures, with its components $n_1 = \sin\alpha\sin\beta$, $n_2 = \sin\alpha\cos\beta$, and $n_3 = \cos\alpha$.

Thus, E_c can be represented as

$$E_c = \frac{g\pi}{12\nu} \sum_{i=1}^{n} \sum_{j=1}^{m_i} W_{ij} r_{ij}^2 b_{ij}^{*3} \overline{J}^{T} (\delta - n_i \otimes n_i)\overline{J}$$

(46)

From Eqs. (39)-(41), (46) and (20), it can be referred that

$$\widetilde{K} = \overline{K}_r + \frac{g\pi}{12\nu V_p} \sum_{i=1}^{n} \sum_{j=1}^{m_i} W_{ij} f^3(\beta_{ij}) r_{ij}^2 b_{0ij}^3 (\delta - n_i \otimes n_i)$$

(47)

In Eq. (47), n is determined by the orientation of the fractures, which reflects the effect of the orientation of the fractures on the fluid flow. r and b_0 represent the size or the scale of the fractures; they retrain the fluid flow through the fractures from their developing magnitude. W is a parameter introduced to show the impact of the connectivity of the fracture network on fluid flow. Finally, f(β) is a function used to demonstrate the coupling effect between fluid flow and stress state. The hydraulic tensor for fractured rock masses given in Eq. (47) is related to the volume of the flow region, V_p, which exactly shows the size effect of the hydraulic properties. Intuitively, the smaller the V_p size is, the less number of fractures is contained within the volume, and thus the poorer the representative of the computed hydraulic conductivity tensor. On the other hand, when V_p is increased up to a certain value, the fractures involved in the cubic volume are dense enough and the hydraulic conductivity tensor for the rock mass does not vary with the size of the volume. This V_p size is exactly the representative elementary volume, REV, of the flow region. The V_p size of the flow region is required to be larger than REV for estimating the hydraulic conductivity tensor for the fractured rock mass. Otherwise, treating the fractured rock mass as an equivalent continuum medium is not appropriate, and the discrete fracture flow approach is preferable.

Comparison with Snow's and Oda's Models

Now we make a comparison between the formulation of the hydraulic conductivity tensor presented in Eq. (47) and the formulation given by Snow (1969) as well as the formulation given by Oda (1986). The Snow's formulation is as follows:

$$\mathbf{K} = \frac{g}{12\nu} \sum_{i=1}^{n} \frac{b_i^3}{s_i} (\boldsymbol{\delta} - \mathbf{n}_i \otimes \mathbf{n}_i) \tag{48}$$

where s_i is the average spacing of the ith set of fractures. If we neglect the hydraulic conductivity of the rock matrix and the connectivity of the factures, and define

$$b_i = \frac{1}{m_i} \sum_{j=1}^{m_i} f(\beta_{ij}) b_{0ij} \quad \text{and} \quad s_i^{-1} = \frac{\pi}{V_p} \sum_{j=1}^{m_i} r_{ij}^2 \tag{49}$$

Then, the formulation presented in Eq. (47) is totally equivalent to Snow's formulation, Eq. (48).

On the other hand, the Oda's formulation is described by

$$\mathbf{K} = \varsigma(P_{kk}\boldsymbol{\delta} - \mathbf{P}) \tag{50}$$

where P is the fracture geometry tensor, with $P_{kk} = P_{11} + P_{22} + P_{33}$.

$$\mathbf{P} = \pi\rho \int_0^{\infty} \int_0^{\infty} \int_{\Omega} r^2 b^3 \mathbf{n} \otimes \mathbf{n} E(n, r, b) d\Omega dr db \tag{51}$$

where E(n, r, b) is a probability density function of the geometry of the fractures, ρ is the number of fracture centers per unit of volume, with $\rho = m_v / V_p$, $m_v = \sum m_i$, and ς is the dimensionless scalar adopted to penalize the permeability of real fractures with roughness and asperities. Assuming that a statistically valid REV exists and being aware that the fracture orientation is a discrete event, the fracture geometry tensor may be empirically constructed by the following direct summation

$$\mathbf{P} = \frac{\pi}{V_p} \sum_{i=1}^{m_v} r_i^2 b_i^3 \mathbf{n}_i \otimes \mathbf{n}_i \tag{52}$$

Following a similar deduction, it can be inferred that all these three formulations are equivalent not only in form but also in function, though they are derived from different approaches and different assumptions. The formulation presented in Eq. (47) can be directly obtained from Snow's formulation by considering the connectivity and roughness of the fractures and integrating the aperture changes under engineering disturbance. The discretized form of the Oda's formulation is much closer to the current formulation, and the latter can also be directly achieved from the former by considering the connectivity of the fracture network. However, the proposed method for formulating an equivalent hydraulic conductivity tensor for complex rock mass based on the superposition principle of liquid dissipation energy is a widely applicable approach not only to equivalent continuum but also to discrete medium.

A Numerical Example: Hydraulic Conductivity of the Rock Mass in the Laxiwa Hydropower Project

In order to validate the theoretical model presented in Section 4.2, we investigated the hydraulic conductivity of a fractured rock mass at the construction site of the Laxiwa Hydropower Project, the second largest hydropower project on the upstream of the Yellow River. The selected construction site for a double curvature arch dam is a V-shaped valley formed by granite rocks, as shown in Fig. 14. The dam height is 250 m, the top elevation of the dam is 2460 m, the reservoir storage capacity is 1.06 billion m³ and the total installed capacity is 4200 MW.

A typical section of the Laxiwa dam site is illustrated in Fig. 15. Besides faults, four sets of critically oriented fractures are developed in the rock mass at the construction site. The geological characteristics of the fractures are described by spacing, trace length, aperture, azimuth, dip angle, the joint roughness coefficient, JRC, of the fractures as well as the connectivity of the fracture network (i.e., the number of fractures intersected by one fracture). According to site investigation, the statistics (i.e., the averages and the mean squared deviations, as well as the distribution of the characteristics) of the fractured rock mass on the right bank of the valley are listed in Table 3.

Figure 14. Site photograph of the Laxiwa valley.

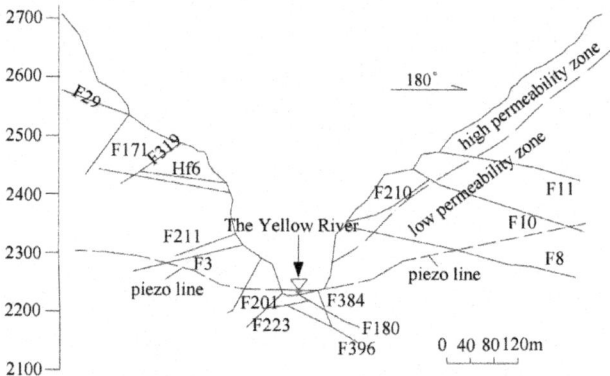

Figure 15. A typical section of the Laxiwa dam site.

Table 3. Characteristic variables of the fractured rock mass[*]

Set	Spacing (m)	Length (m)		Aperture (mm)		Azimuth (°)		Dip (°)		Connectivity	
		avg.	dev.	avg.	dev.	avg.	dev.	avg.	dev.	avg.	dev.
1	1.45	5	1.5	0.096	0.02	85.3	10	54.5	10	5	3
2	2.62	3	1.0	0.096	0.02	355.1	20	29.8	5	3	2
3	10.96	3	1.0	0.096	0.02	287.4	20	61.4	10	3	2
4	10.96	3	1.0	0.096	0.02	320.2	20	11.9	5	3	2
Distribution	logarithmic normal	negative exponential		Gama		normal		normal		normal	

[*]'avg.' denotes arithmetic mean of a variable,
'dev.' represents root mean squared deviation

At the construction site of the Laxiwa dam, a total number of 1450 single-hole packer tests were conducted to measure the hydraulic properties of the rock mass, with 113 packer tests for the shallow rock mass on the right bank in 0−80 m horizontal depth and 278 packer tests for the deeper rock mass. The measurements of the hydraulic conductivity range from 10^{-5} cm/s to 10^{-6} cm/s for the shallow rock mass and from 10^{-6} cm/s to 10^{-7} cm/s for the deeper rock mass, with in average 4.94×10^{-5} cm/s for the former and 3.80×10^{-6} cm/s for the latter, respectively (Liu, 1996). On the other hand, in-situ stress tests showed that the geostress in the base of the valley and in deep rock mass has a magnitude of 20−60 MPa, with the direction of the major principal stress pointing towards NNE. As a result of stress release, the release fractures are frequently developed and a high permeability zone of 0−80 m horizontal depth is formed in the bank slope, as shown in Fig. 15. The stress release fractures, however, become infrequent in deeper rock mass, and the measured hydraulic conductivity is generally 1−2 orders of magnitude smaller than the hydraulic conductivity of the rock mass in shallow depth away from the bank slope. Therefore, the hydraulic conductivity of the rock mass at the construction site of the Laxiwa arch dam is mainly controlled by the fracture network and the stress state.

Based on these statistics given in Table 3, fracture networks can be generated and calibrated for the rock mass at the construction site of the Laxiwa Hydropower Project using the MonteCarlo method by assuming that each fracture is a smooth, planar disc, with its center uniformly distributed in the simulated area. For each set of fractures, the geometrical parameters of any one are sampled by Monte-Carlo method until enough fractures are included in the simulated area. Then, a calibration procedure is invoked to check whether the generated model satisfies the distribution mode of the real fracture network. If doesn't, the fracture network will be regenerated until one matches the distribution mode. With the generated fracture network, the actual connectivity can be computed by spatial operation on the fractures. But for calibrated fracture network, a more convenient approximate approach to determine the connectivity of the fracture network, as it is adopted here, is to directly produce ξ_{ij} in Eq. (42) with the Monte-Carlo method and the characteristics presented in Table 3, then W_{ij} is derived from Eq. (42) with $\bar{\xi}_i$, the maximum number of fractures cut by the ith set of fractures. Field measurements are used to estimate $\bar{\xi}_i$, with $\bar{\xi}_1 = 11$, $\bar{\xi}_2 = 8$ and $\bar{\xi}_3 = \bar{\xi}_4 = 6$ for the four sets of fractures, respectively. Fig. 16 illustrates a simulated fracture network with size of $20 \times 20 \times 20$ m.

On the basis of the fracture network generated above, we compute the hydraulic conductivity tensor for the simulated cubic volume of rock mass with size of 20×20×20 m using the method given by Snow (1969) and the method presented in Section 4.2, respectively. To show the coupling effect of stress/deformation on hydraulic properties, we consider two scenarios for examination. In the first scenario, we consider the fracture network located in the shallow depth away from the bank slope, where the impact of the in-situ stress is negligible. While in the second scenario, the fracture network is situated in larger depth, and a typical stress state with $\sigma_x=\sigma_z=10$ MPa and $\sigma_y=20$ MPa is associated with it. Based on laboratory test results, the shear modulus of the fractures is estimated as $\mu=2$ MPa, and then by taking the Poisson's ratio as $v=0.25$, the Lame's constant is derived with $\lambda=2$ MPa. The kinematic viscosity of underground water is set to be $v_w=1.14\times10^{-6}$ m2/s and the frictional angle-like parameter and the normal stress-like parameter are taken as $\varphi=0.4363$ and $s=\sigma_n/20$.

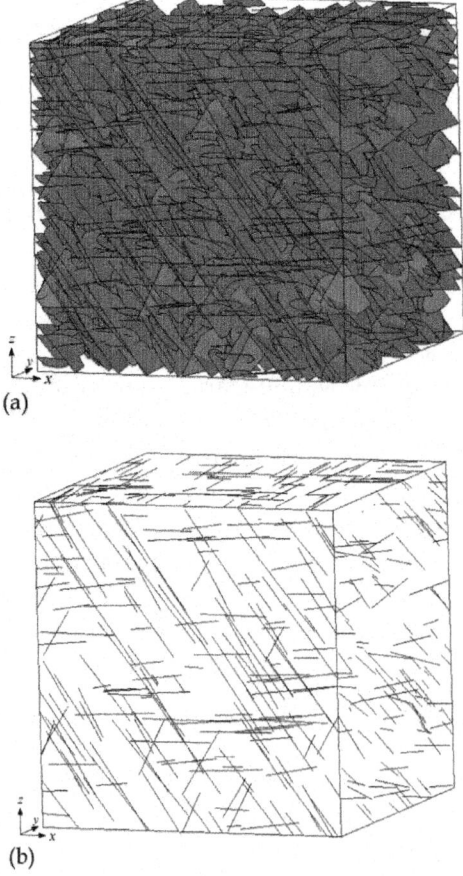

(a)

(b)

Figure 16. A three dimensional fracture network with size of 20×20×20 m generated by using the Monte-Carlo method for the rock mass in the Laxiwa Hydropower Project: (a) fracture network and (b) traces of the fractures on the surfaces of the simulated area.

The predicted hydraulic conductivity tensor for the examined rock mass is listed in Table 4. From Table 4, one observes that for shallow rock mass (where the effect of in-situ stress is not considered), the Snow's method and the method presented in Section 4.2 predict similar results and the predicted hydraulic conductivity is in the magnitude of 10^{-5} cm/s and close to in-situ hydraulic observations, but the anisotropy in hydraulic conductivity manifests due to non-uniform distribution of fractures. Compared with the hydraulic conductivity of the shallow rock mass, the predicted hydraulic conductivity for the rock mass in larger depth with the same fracture network decreases in 2 orders of magnitude due to the closure of the fractures applied by the in-situ stresses, but the anisotropic property of the hydraulic conductivity remains, which suggests that the occurrence of the fractures has a significant impact on permeability. Taking into consideration the applied stress level, the reduction of hydraulic conductivity in orders of magnitude is very close to the results achieved in Min et al. (2004) through a discrete element method, and generally agrees with the in-situ hydraulic observations.

Table 4. Predicted hydraulic conductivity tensor of the rock mass at the construction site of the Laxiwa dam (cm/s)

Snow's model (for shallow rock mass)		
4.78E–05	–4.76E–07	–1.71E–05
–4.76E–07	7.49E–05	–1.41E–05
–1.71E–05	–1.41E–05	4.08E–05
The proposed model (for shallow rock mass)		
1.93E–05	–1.75E–07	–6.39E–06
–1.75E–07	2.99F–05	–5.81E–06
–6.39E–06	–5.81E–06	1.64E–05
The proposed model (for deep rock mass)		
9.06E–08	–4.81E–09	–6.10E–08
–4.81E–09	1.85E–07	–1.92E–08
–6.10E–08	–1.92E–08	1.10E–07

Now, we take for example the rock mass in shallow depth to estimate the REV size of the rock mass. For this purpose, the scale of the rock mass is increased gradually from 3×3×3 m to 40×40×40 m with an increment of 1 m in each dimension. In each step, a fracture network with prescribed size is generated by using the Monte-Carlo method described above, and it is worth noting that this method is somewhat different from the method used by Min &

Jing (2003) and Long et al. (1982). For each fracture network, the hydraulic conductivity tensor is calculated from Eq. (47) and then the principal hydraulic conductivities are further obtained from the hydraulic conductivity tensor. The relationship between the computed principal hydraulic conductivities and the sizes of the rock mass is illustrated in Fig. 17. As we can see from Fig. 17, when the block size of the rock mass is smaller than 18×18×18 m, the population of fractures is not dense enough and the principal hydraulic conductivities fluctuate dramatically. On the other hand, as the size scales up to about 20×20×20 m, the examined rock mass has included enough fractures and the computed principal hydraulic conductivities approach a rather steady state, with k_1, k_2, k_3 estimated to be $2.41×10^{-5}$ cm/s, $3.59×10^{-5}$ cm/s, $1.08×10^{-5}$ cm/s, respectively. This suggests that the REV does exist in the rock mass and the rock mass can be regarded as an equivalent continuum medium as long as its size is no less than, e.g., 20×20×20 m or 8000 m³.

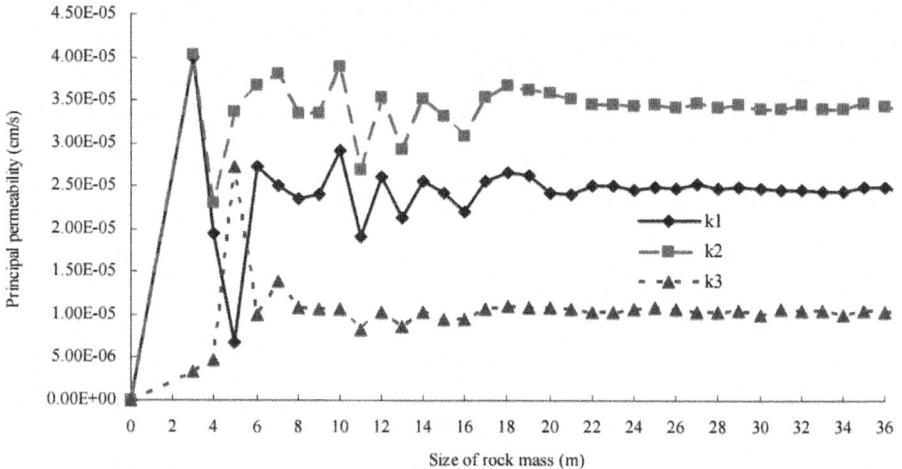

Figure 17. Hydraulic conductivity versus the volume size of the fractured rock mass.

STRAIN-DEPENDENT HYDRAULIC CONDUCTIVITY TENSOR OF FRACTURED ROCKS

On the basis of the strain-dependent model presented in Section 3 for rock fractures, this section formulates the strain-dependent hydraulic conductivity tensor for fractured rock masses cut by one or multiple sets of parallel fractures. The major difference between the model in this section and the stress-dependent model presented in Section 4 is that the former is capable of describing influence of the post-peak mechanical behaviors on the hydraulic properties of the rock masses, and is suited for modelling the coupled processes in rock masses at high stress level and in drastic engineering disturbance condition.

An Equivalent Elasto-Plastic Constitutive Model for Fractured Rocks

Consider a fractured rock mass cut by n sets of planar and parallel fractures of constant apertures with various orientations, scales and densities. The global response of the fractured rock mass under loading comes both from weak fractures and from stronger rock matrix. Based on this observation, an equivalent elasto-plastic constitutive model can be formulated by imposing assumptions on the interaction between fractures and rock matrix. The coordinate systems are defined in the same way with those defined in Section 4.1 (see Fig. 13). Denote the unit vector along X_i-axis of the global frame as e_i (i=1, 2, 3) and the unit vector along x_i^f-axis of the fth local frame as e_i^f (i=1, 2, 3). Then, a second order tensor, $l f$, can be defined for transforming physical quantities between the frames, with the components in the form of

$$l_{ij}^f = e_i^f \cdot e_j$$

$$(53)$$

Regarding the fractured rock mass as a continuous medium at the macroscopic scale, it is rational to assume that the global strain increment of the fractured rock mass is composed of the strain increments of rock matrix and fractures (Pande & Xiong, 1982; Chen & Egger, 1999), i.e.

$$d\varepsilon = d\varepsilon^R + \sum_F d\varepsilon^F$$

$$(54)$$

where $d\varepsilon$, $d\varepsilon^R$ and $d\varepsilon^F$ are the total incremental strain tensor, the incremental strain tensor of rock matrix and the incremental strain tensor of fth set of fractures measured in the global coordinate system, respectively. Note that a variable with a superscript in upper case (i.e. R or F) means that it is measured in the $X_1X_2X_3$ system, while a variable with a superscript in lower case (i.e. f) is measured in $x_1^f x_2^f x_3^f$ system, respectively. Unless otherwise specified, the superscripts F and f are not summing indices.

On the other hand, traction continuity has to be ensured across the fracture interfaces. In the global coordinate system, this condition can be strictly represented by (Pande & Xiong, 1982; Chen & Egger, 1999)

$$d\sigma = d\sigma'^R = d\sigma'^F \qquad (55)$$

where $d\sigma'$, $d\sigma'^R$ and $d\sigma'^F$ are the effective incremental stress tensor of the fractured rock mass, the effective incremental stress tensor of rock matrix and the effective incremental stress tensor of fth set of fractures, respectively. The effective stress tensor σ' is defined as

$$\boldsymbol{\sigma}' = \boldsymbol{\sigma} + \alpha p \boldsymbol{\delta} \tag{56}$$

where σ is the total stress tensor (positive for tension), p is the pore water pressure (positive for compressive pressure), and α (α≤1) is an effective stress parameter.

Combining the plastic potential flow theory and the consistency conditions of rock matrix and fractures, an equivalent elasto-plastic constitutive model can be derived from Eqs. (54) and (55):

$$d\boldsymbol{\varepsilon} = \mathbf{S}^{ep}:d\boldsymbol{\sigma}' \tag{57}$$

with

$$\mathbf{S}^{ep} = (\mathbf{C}^{R,ep})^{-1} + \sum_{F}(\mathbf{C}^{F,ep})^{-1} \tag{58}$$

where S^{ep} is the equivalent elasto-plastic compliance tensor of the fractured rock mass.

$C^{R,ep}$ in Eq. (58) is the elasto-plastic modulus tensor of rock matrix. Neglecting the degradation of rock strength in the volume close to fracture intersections, $C^{R,ep}$ can be written as

$$\mathbf{C}^{R,ep} = \mathbf{C}^{R} - \frac{\mathbf{C}^{R}:\dfrac{\partial Q_R}{\partial \boldsymbol{\sigma}'} \otimes \dfrac{\partial F_R}{\partial \boldsymbol{\sigma}'}:\mathbf{C}^{R}}{\dfrac{\partial F_R}{\partial \boldsymbol{\sigma}'}:\mathbf{C}^{R}:\dfrac{\partial Q_R}{\partial \boldsymbol{\sigma}'} + H_R} \tag{59}$$

where C^R is the fourth-order elastic modulus tensor of rock matrix, which can be represented in terms of the Lame's constants λ and μ:

$$C_{ijkl}^{R} = \lambda \delta_{ij}\delta_{kl} + \mu(\delta_{ik}\delta_{jl} + \delta_{il}\delta_{jk}) \tag{60}$$

F_R, Q_R and H_R in Eq. (59) are the yield function, the plastic potential function and the hardening modulus of rock matrix, respectively. A non-associative flow rule with elastic perfectly plasticity (i.e. H_R=0) is adopted for better modeling dilatant behavior of rock matrix by virtue of, for example, the Druker-Prager criterion with its cone fully inscribed by the Mohr-Coulomb hexagon, defined by functions

$$F_R = aI_1' + \sqrt{J_2} - \kappa = 0 \tag{61}$$

$$Q_R = \beta I_1' + \sqrt{J_2} \tag{62}$$

with

$$\alpha = \sin\varphi_R \, / \, \sqrt{3(3 + \sin^2\varphi_R)} \tag{63}$$

$$\kappa = 3c_R \cos\varphi_R \, / \, \sqrt{3(3 + \sin^2\varphi_R)} \tag{64}$$

$$\beta = \sin\psi_R \, / \, \sqrt{3(3 + \sin^2\psi_R)} \tag{65}$$

where c_R and φ_R are the cohesion and the friction angle of rock matrix, respectively. I_1' and J_2 are the first invariant of the effective stress and the second invariant of the deviatoric stress of rock matrix, respectively. ψ_R is the mobilized dilatancy angle of rock matrix.

It should be noted here that in the literature, Drucker-Prager criterion has been used by many authors to model the elasto-plastic behavior of intact rock matrix, see Pande & Xiong (1982) and Chen & Egger (1999) for example. Although a modified Drucker-Prager yield function may be more suitable for this formulation in order to model plastic deformation properties of intact rock such as pressure dependency, strain hardening, transition from compressibility to dilatancy and stress path dependency (Chiarelli et al., 2003), the criterion given above may keep the formulation compact and does not lose generality. Other yield functions, such as the modified Drucker-Prager criterion (Chiarelli et al., 2003) or the modified Hoek-Brown criterion (Hoek et al., 1992), can also be integrated into the formulation without major mathematical difficulties.

With the researches conducted by Yuan & Harrison (2004) and Alejano & Alonso (2005), the decaying process of the rock dilatancy angle in line with plasticity can be described by the following negative exponential expression through the equivalent plastic strain of rock matrix, $\bar{\varepsilon}_R^p$ (Lai, 2002):

$$\psi_R = \psi_R^{peak} \exp(-r_R \bar{\varepsilon}_R^p) \tag{66}$$

where $r_R \geq 0$ is a parameter for modelling the decaying process of the dilatancy angle, and ψ_R^{peak} is the peak dilatancy angle of rock matrix and the following expression has been proposed by recovering the shape of the peak dilatancy angle of fractures given by Barton & Bandis (1982) and by assuming $\psi_R^{peak} = \varphi_R$ for null confinement pressures (Alejano & Alonso, 2005):

$$\psi_R^{peak} = \frac{\varphi_R}{1 + \log_{10}\sigma_c} \log_{10}\frac{\sigma_c}{-\sigma_3' + 0.1} \tag{67}$$

where σ_c is the unconfined compressive strength for intact rock. By Eqs. (66) and (67), the dependencies of rock dilatancy on plasticity, confining stress and scale are produced.

The equivalent plastic strain $\bar{\varepsilon}^P$ is computed by the following:

$$\bar{\varepsilon}^P = \int d\bar{\varepsilon}^P = \int \sqrt{\frac{2}{3} d\varepsilon^P : d\varepsilon^P}$$

(68)

Similarly, $C^{F,ep}$ in Eq. (58) is the elasto-plastic modulus tensor of fth set of fractures measured in the $X_1 X_2 X_3$ system, which can be calculated from its corresponding elastoplastic modulus tensor measured in the $x_1^f x_2^f x_3^f$ system, $C^{f,ep}$, with the assumption of small strain and by imposing the following tensor transformation:

$$C_{ijkl}^{F,ep} = l_{mi}^f l_{nj}^f l_{ok}^f l_{pl}^f C_{mnop}^{f,ep}$$

(69)

with

$$C^{f,ep} = C^f - \frac{C^f : \dfrac{\partial Q_f}{\partial \sigma'} \otimes \dfrac{\partial F_f}{\partial \sigma'} : C^f}{\dfrac{\partial F_f}{\partial \sigma'} : C^f : \dfrac{\partial Q_f}{\partial \sigma'} + H_f}$$

(70)

where Cf is the fourth-order tangential elastic modulus tensor of the fth set of fractures, with $C_{3333}^f = s_f k_{nf}$, $C_{2323}^f = C_{3131}^f = s_f k_{sf}$, and with all other elements equal to zero. The symbols k_{nf}, k_{sf} and s_f are the normal stiffness, the tangential stiffness and the spacing of the fth set of fractures, respectively. The expressions for the elements in Cf mean that the strain of fractures is evaluated over the fracture spacing, not over the fracture aperture, thus enabling the proposed model to consider the post-sliding plasticity of fractures and nonlinear variations of k_{nf} and k_{sf} with dilatancy caused by shear loading, without violating the small strain assumption.

F_f, Q_f and H_f in Eq. (70) are the yield function, the plastic potential function and the hardening modulus of the fth set of fractures, respectively. The elasto-plastic behavior of the fractures is treated in a similar fashion as that for the rock matrix, with a non-associative Mohr-Coulomb criterion:

$$F_f = \sqrt{\tau_{zxf}^2 + \tau_{zyf}^2} + \sigma_{zf}' \tan\varphi_f - c_f = 0$$

(71)

$$Q_f = \sqrt{\tau_{zxf}^2 + \tau_{zyf}^2} + \sigma_{zf}' \tan\psi_f$$

(72)

where σ_{zf}', τ_{zxf} and τ_{zyf} are the effective normal stress and the shear stresses on the fracture surfaces, respectively. c_f, φ_f and ψ_f are the cohesion, the friction angle and the mobilized dilatancy angle of the fth set of fractures, respectively. Similar to Eq. (66), ψ_f is also a shrinking function of the equivalent plastic strain of fractures $\bar{\varepsilon}_f^P$, and depends on normal stress and scale as well, in the following form:

$$\psi_f = \psi_f^{\text{peak}} \exp(-r_f \bar{\varepsilon}_f^P)$$

(73)

where r_f is the decaying parameter and ψ_f^{peak} is the peak dilatancy angle of the fth set of fractures, respectively, with the latter calculated by Eq. (26).

Thus at any loading step, as long as the stress increment of the equivalent rock mass, $d\sigma'$, is obtained, the local strain pertinent to fth set of fractures can be derived as follows:

$$d\varepsilon^F = (C^{F,ep})^{-1} : d\sigma'$$

(74)

and

$$d\varepsilon_{ij}^f = l_{im}^f l_{jn}^f d\varepsilon_{mn}^F$$

(75)

The separation of the incremental strain of fractures from that of the rock mass through the proposed equivalent constitutive model plays a significant role in the present study. It enables the formulation of strain-dependent hydraulic conductivity that accounts for the mobilized dilatancy behavior, which will be demonstrated in the following section.

Strain-Dependent Hydraulic Conductivity Tensor for Fractured Rocks

Consider a domain of flow that has been discretized into several sub-domains according to rock quality classification. Suppose that each sub-domain contains n sets of fractures, with average initial aperture b_{f0} and spacing s_f for the fth set of fractures. Starting from Eq. (22) and using the averaging concept for the hydraulic conductivity over the whole sub-domain, the equivalent initial hydraulic conductivity of the fth set of fractures, k_{f0}, in the examined sub-domain can be represented as (Castillo, 1972; Liu et al., 1999)

$$k_{f0} = \varsigma \frac{g b_{f0}^3}{v s_f}$$

(76)

where ς, as pointed out before, is a dimensionless constant introduced to penalize the real water conducting capacity of natural fractures with rough walls, finite scales, asperity areas and filling materials. The validity of using a constant value of ς has been examined by Zhou et al. (2006).

Assuming that the change in spacing s_f during modeling is negligible, under normal and shear stress loadings we have

$$k_f = \varsigma \frac{gb_f^3}{vs_f} = \varsigma \frac{g(b_{f0} + \Delta b_f)^3}{vs_f}$$
(77)

where Δb_f and k_f are the increment of the aperture and the equivalent hydraulic conductivity of the fth set of fractures under loading, respectively. Suppose that strain localization (Lai, 2002; Vajdova, 2003) is not dominantly exhibited in the concerned fractures, it is approximately valid that

$$\Delta b_f = s_f \Delta \varepsilon_{zf}$$
(78)

where $\Delta \varepsilon_{zf}$ is the increment of the normal strain of the fth set of fractures, which can be directly obtained from Eq. (75).

Substituting Eq. (78) into Eq. (77) then yields

$$k_f = k_{f0} \left(1 + \frac{s_f}{b_{f0}} \Delta \varepsilon_{zf} \right)^3$$
(79)

Following the theory proposed by Snow (1969), a strain-dependent equivalent hydraulic conductivity tensor for fractured rock masses with n sets of fractures is represented by

$$\mathbf{K} = \sum_f k_f (\boldsymbol{\delta} - \mathbf{n}_f \otimes \mathbf{n}_f) = \sum_f k_{f0} \left(1 + \frac{s_f}{b_{f0}} \Delta \varepsilon_{zf} \right)^3 (\boldsymbol{\delta} - \mathbf{n}_f \otimes \mathbf{n}_f)$$
(80)

where K is the equivalent hydraulic conductivity tensor of the examined rock mass, and n_f is the unit vector normal to the fth set of fractures.

The following significant implications can be observed from the formulation of K in Eq. (80):

- K is a cubic function of $\Delta \varepsilon_{zf}$, and any variation in ε_{zf} under loading will trigger the change in K, even in orders of magnitude. This exactly accounts for the coupling effect of mechanical loading (strain/stress) on hydraulic properties.

- K depends on incremental strains, rather than on stresses, which makes it possible to integrate various material nonlinearities in hydro-mechanical coupling analysis.

- In addition to cubic relation, the influence of $\Delta\varepsilon_{zf}$ on K is amplified by s_f/b_{f0}, indicating that K can be rather sensitive to b_{f0} and s_f. Therefore, techniques for estimating b_{f0} and s_f need to be carefully developed, on the basis of laboratory or in-situ hydraulic test data.

- The orientations of fractures possibly render K highly anisotropic, even if K is initially assumed isotropic, as has been systematically examined, e.g. by Liu et al. (1999).

- When implemented in a FEM code, a different K can be associated to each geological sub-domain or even to each element, as long as k_{f0}, b_{f0} and s_f for the sub-domains or elements can be estimated in advance.

- As a nature of the homogenized equivalent continuum approach, the size effect of fractures, especially the size-dependency of aperture, is not fully considered in the formulation of K for simplicity, even though it can be reflected to some degree through ς and scaled JRC and JCS values. The connectivity and the intersection effect of fractures, on the other hand, may have a more significant influence on K, but similarly, they cannot be properly considered in the equivalent continua without explicit representation of fractures. A rough remedy is to process the fracture system in such a way that only the connected fracture populations are included for conducting analyses.

To determine K of a fractured rock under any loading paths, a coupled hydro-mechanical process has to be invoked. With the assumption of incompressible rock matrix and fluid (e.g. groundwater), the governing equations for the coupled process of saturated fluid flow and deformation are given below as balance equation, geometric equation and fluid flow equation, respectively:

$$\sigma'_{ij,j} - \alpha p_{,i} + f_i = 0 \tag{81}$$

$$\varepsilon_{ij} = \frac{1}{2}\left(u_{i,j} + u_{j,i}\right) \tag{82}$$

$$\frac{\partial}{\partial x_i}\left(k_{ij}\frac{\partial h}{\partial x_j}\right) = \frac{\partial \varepsilon_v}{\partial t} \tag{83}$$

where f_i and u_i are the components of the body force and displacement in the ith direction, $h=p/\gamma_w+z$ the water head, z the vertical coordinate, γ_w the unit weight of water, and ε_v the volume strain of the rock mass.

In the coupled process given above, mechanical loading or disturbance to the rock mass results in change in flow properties and flow behavior through Eqs. (80) and (83), while the change in flow behavior leads to change in mechanical response of the rock mass through Eq. (81). When the coupled process reaches a stable state, the solution to K is also available.

Now we briefly discuss how to determine k_{f0}, b_{f0} and s_f in Eq. (80) based on laboratory or insitu hydraulic test or site investigation data. Obviously, the initial hydraulic conductivity, k_{f0}, can be determined by in-situ hydraulic tests. Suppose the initial hydraulic conductivity tensor, K_0, is known through in-situ hydraulic test, as suggested by Hsieh & Neuman (1985), then K_0 can be rewritten, from Eq. (80), in the following form:

$$K_0 = \sum_f k_{f0}(\delta - n_f \otimes n_f)$$

(84)

By optimizing Eq. (84), k_{f0} (f=1, ..., n) can be estimated if the number of the sets of critically oriented fractures, n, is less than or equal to 6 (i.e. the number of the independent components of K_0), regardless K_0 is assumed to be isotropic or anisotropic.

The average spacing of the fth set of fractures, s_f, can be roughly estimated from the statistics of drill holes or scanlines. An alterative, however, is to use RQD (Rock Quality Designation) for determining s_f, as suggested by Liu et al. (1999), when the value of RQD for a specific rock mass is known a priori.

After the initial hydraulic conductivity, k_{f0}, and the average spacing, s_f, of the fractures are determined, the mean initial aperture of the fractures, b_{f0}, is ready to be back-calculated from Eq. (76).

Validation of the Proposed Model

Hydraulic Conductivity of the Surrounding Rock of a Circular Tunnel in the Stripa Mine

Here we compare the proposed method with results from a previous study as presented by Liu el al. (1999) by applying the method to an excavated circular tunnel with a biaxial stress field, σ_x and σ_z. The physical model is illustrated in Fig. 18, which is actually a manifestation of the reality of the Stripa mine in Sweden (Kelsall et al., 1984; Pusch, 1989). The following description about the tunnel is directly taken from Liu et al. (1999):

A Buffer Mass Test was conducted in Stripa Mine over the period 1981-1985 (Kelsall et al., 1984; Pusch, 1989) to measure the permeability of a large volume of low permeability fractured rock mass by monitoring water flow into a 33 m long section of the tunnel, as a large scale in-situ experiment for the research and development programs of underground geological disposal of nuclear wastes of the participating countries of the Stripa Project. The radius of the tunnel is about 2.5 m with two major sets of fractures striking obliquely to the tunnel axis, as shown in Fig. 18.

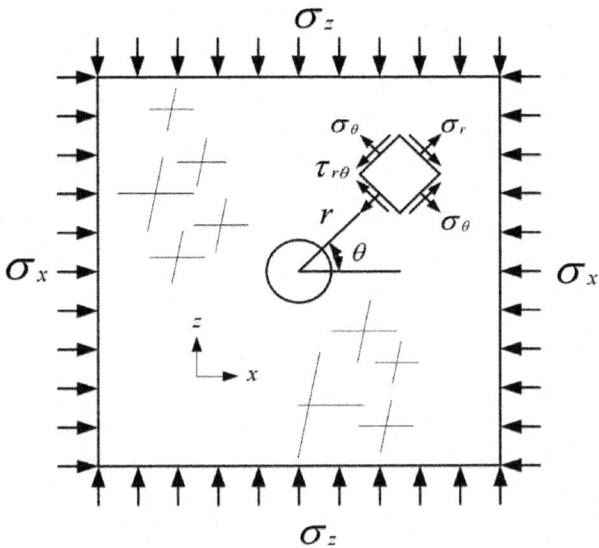

Figure 18. Sketch of a circular excavation in a biaxial stressed rock mass.

Fracture frequency measured in holes drilled from the tunnel was on average 4.5 fractures/m in inclined holes and 2.9 fractures/m in vertical holes. The initial stress field is anisotropic with high horizontal stress component and the conductivity of the virgin rock is about 10^{-10} m/s. The excavation of the test drift produced a dramatic increase in axial hydraulic conductivity in a narrow zone adjacent to the periphery of the drift. The conductivity increase is estimated to be 3 orders of magnitude.

The following assumptions are made in the calculations, with some of them similar to those in Liu et al. (1999):

- Statically uniform aperture and spacing distributions exist before excavation;
- Fracture spacing and continuity are not altered by the excavation;
- The high obliquity of the two major sets of fractures can be well approximated by two orthogonal sets of fractures;
- Excavation-induced strain redistribution may be adequately captured by the proposed equivalent elasto-plastic constitutive model.

Some of the parameters are directly taken from Liu et al. (1999), while other unavailable parameters are assumed, as listed in Table 5, in which the initial mechanical aperture of the fractures is back-calculated from Eq. (76)

by taking $k_0=10^{-10}$ m/s. Consistent with Liu et al. (1999), the far-field stress components are taken as $\sigma_x=20$ MPa and $\sigma_z=10$ MPa, respectively.

Table 5. Geometrical and mechanical parameters for a circular tunnel

Category	Parameter	Setting
Intact rock matrix	Elastic modulus, E	37.5 GPa
	Poisson's ratio, v	0.25
	Cohesion, c_R	5 MPa
	Friction angle, φ_R	46°
Fractures	Initial mechanical aperture, b_0	0.0075 mm
	Spacing, s	0.27 m
	Normal stiffness, k_n	200 GPa/m
	Shear stiffness, k_s	100 GPa/m
	Dimensionless constant, ς	0.0067
	Cohesion, c_f	0.4 MPa
	Friction angle, φ_f	40°

To avoid the difficulty in determining the initial dilatancy angles and the corresponding decay parameters of fractures and intact rock matrix, associative flow rule is used in this simulation. Again for simplicity, both the normal stiffness and the shear stiffness of the fractures are assumed constant during excavation. The finite element mesh of the model is shown in Fig. 19, and the FEM program was run to simulate the excavation effect of the tunnel. Fig. 20 shows the deformation zone and plastic zone of the rock mass after the tunnel excavation. Fig. 21 plots the excavation-induced changes in hydraulic conductivities around the circular tunnel, which are directly compared with the results presented in Liu et al. (1999).

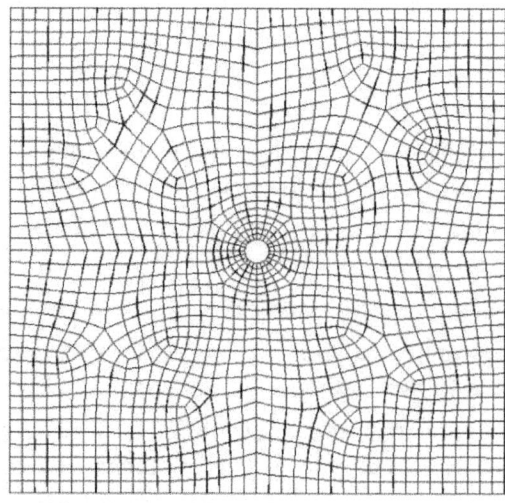

Figure 19. Finite element mesh for simulation of a tunnel excavation.

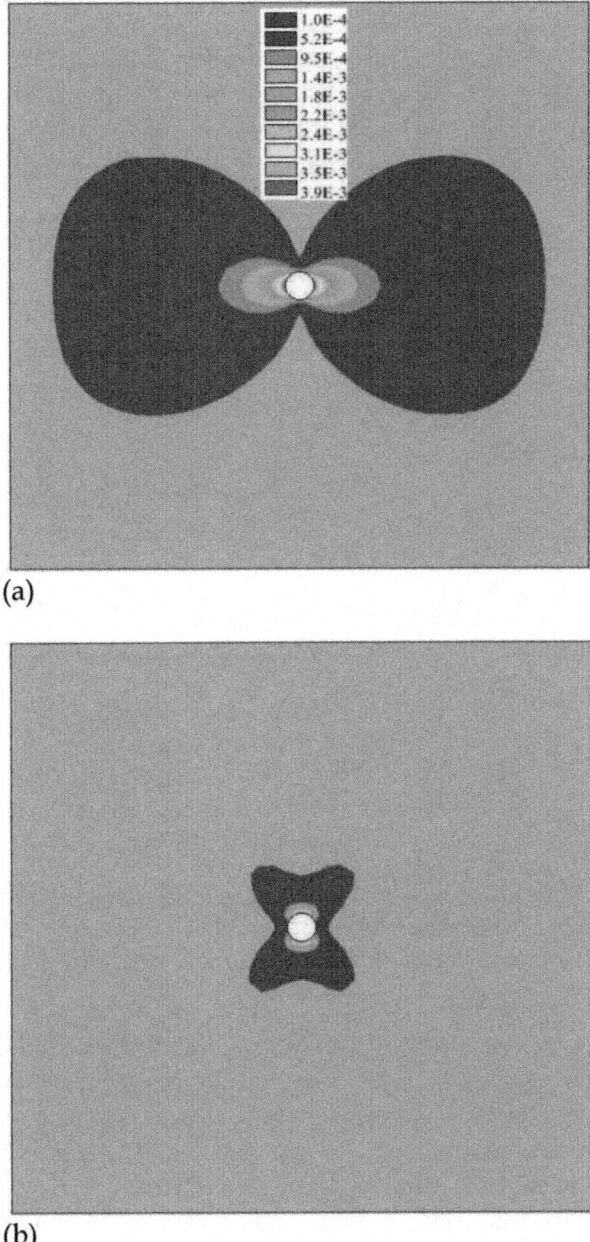

(a)

(b)

Figure 20. Deformation zone and plastic zone induced by the tunnel excavation: (a) deformation zone and (b) plastic zone.

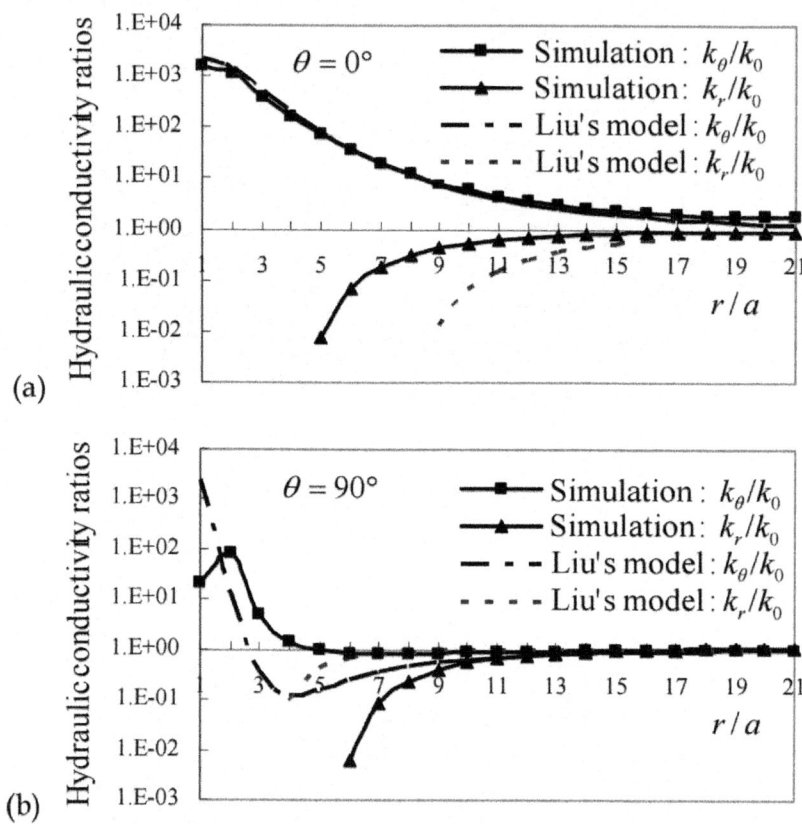

Figure 21. Excavation-induced hydraulic conductivity ratios around a circular tunnel in a biaxial stressed rock mass, where a is the radius of the tunnel and r is the distance away from the tunnel center. θ=0° denotes the horizontal direction while θ=90° the vertical direction.

It can be observed from Fig. 21 that generally tangential conductivities are found to increase greatly due to the formation of the excavation disturbed zone around the tunnel, while radial conductivities diminish greatly as a result of closure on related fractures. In the horizontal direction (i.e. θ=0°), the excavation-induced tangential hydraulic conductivity ratios, k_θ/k_0, predicted by our model are very close to the results presented in Liu et al. (1999). For radial hydraulic conductivity ratios, k_r/k_0, however, deviation occurs in the vicinity of the excavation. Such a deviation is also found both for k_θ/k_0 and for k_r/k_0 in the vertical direction (i.e. θ=90°).

Clearly, these deviations are largely resulted from the facts that (1) Different strain distribution patterns are assumed in the elastic model in Liu et al. (1999) and in our elastoplastic model; (2) Different methods are used to compute the strain increments of fractures. In Liu et al. (1999), normal strains of fractures were separated from rock matrix through a modulus reduction ratio empirically defined as a function of RMR, while in this simulation fracture strains were calculated by strain decomposition through an equivalent elasto-plastic constitutive model; (3) Radial and tangential fractures were assumed in Liu et al. (1999), leading to different background fracture networks; and (4) As mentioned above, some of the parameters, such as the shear strength of fractures and rock matrix, the shear stiffness and normal stiffness of the fractures, are unavailable in the literature (Kelsall et al., 1984; Pusch, 1989; Liu et al., 1999) and hence are empirically assumed in the calculations. If these parameters are determined based on in-situ or laboratory experiments, more convincing results may be achieved.

Despite the deviations, the trends of variation of the hydraulic conductivity ratios around the tunnel due to excavation are consistent between the two studies, and basically accord with the in-situ experimental observations, demonstrating the applicability of the present model in this section.

From Fig. 20, one observes that the excavation-induce deformation zone and plastic zone are asymmetric, due to the anisotropic initial stress field. As a result, the predicted hydraulic conductivities are highly anisotropic due to strain redistribution, as shown in Fig. 21. In the horizontal direction (i.e. $\theta=0°$), the deformation zone extends as far as more than 16 times of the tunnel radius and the plastic zone extends 2 times of the tunnel radius, while in the vertical direction (i.e. $\theta=90°$), they are, respectively, within 2 and 5 times of the tunnel radius. The asymmetry of deformation zone and plastic zone demonstrates why the predicted hydraulic conductivities approach k_0 more slowly in the horizontal direction than in the vertical direction. The changes in hydraulic conductivities resulted from strain redistribution in the disturbed rock mass indicate that a different hydraulic conductivity tensor should be associated to each geological sub-domain or even each element of the rock mass, which is important for hydro-mechanical coupling analyses.

Hydraulic Conductivity of a Cubic Block of Rock Mass with Three Orthogonal Sets of Identical Fractures

In this section, a numerical simulation is conducted to evaluate hydraulic behavior of a cubic block of rock mass containing three orthogonal sets of identical fractures under isotropic triaxial compression and shear loading. The primary goal is to investigate the change in the hydraulic conductivity of the rock mass with increasing shear load, which is obviously not achievable through any elastic models considering only the deformation of fractures under normal stresses, e.g. in Liu et al. (1999). The underlying rock mass block model for examination, with a size of 10×10×10 m (a scale that can represent both the initial mechanical and hydraulic REVs (Min et al., 2004)), is assumed to contain three orthogonal sets of identical fractures, as sketched in Fig. 22. The spacing, s, of each set of fractures and the initial aperture, b_0, of each fracture are assumed to be identical, with s=1 m and b_0=1 mm. The mechanical properties of each fracture are also regarded identical and for simplicity, both the normal stiffness and the shear stiffness of the fractures are assumed to be constant during shear loading. All parameters used in this simulation are listed in Table 6, and such parameter settings enable us to demonstrate how the hydraulic conductivity evolves from initial isotropy to anisotropy in the shearing process.

The examined rock mass block model is divided into 1000 brick elements, and the resultant mesh is shown in Fig. 22. The loading condition is as follows. First, triaxial compressive stresses are applied on the surfaces of the cubic block, with $\sigma_x=\sigma_y=\sigma_z$=20 MPa. Then, a shearing load , τ, is applied on the upper and lower surfaces of the block model step by step, increasing at an increment of 1 MPa until a maximum shear load, 20 MPa, is reached. At each step of shear loading, numerical divergence may occur. If numerical divergence does occur, the simulation program terminates after 1000 iterations with a modified NewtonRaphson method.

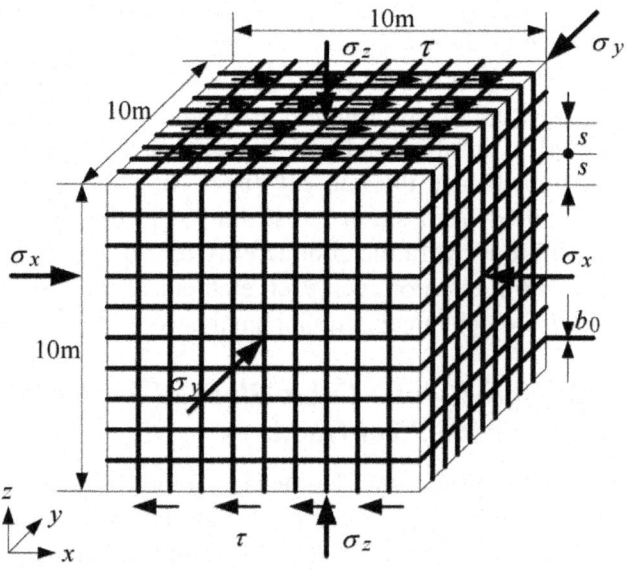

Figure 22. Sketch of a cubic block of rock mass with three orthogonal sets of identical fractures.

Table 6. Geometrical and mechanical parameters for a cubic block of fractured rock mass

Category	Parameter	Setting
Intact rock matrix	Elastic modulus, E	6 GPa
	Poisson's ratio, ν	0.25
	Cohesion, c_R	1 MPa
	Friction angle, φ_R	46°
	Peak dilatancy angle, ψ_R^{peak}	35°
	Decay parameter of dilatancy, r_R	100
Fractures	Initial mechanical aperture, b_0	1 mm
	Spacing, s	1 m
	Normal stiffness, k_n	30 GPa/m
	Shear stiffness, k_s	10 GPa/m
	Dimensionless constant, ς	0.0067
	Cohesion, c_f	0.4 MPa
	Friction angle, φ_f	40°
	Peak dilatancy angle, ψ_f^{peak}	26°
	Decay parameter of dilatancy, r_f	100

Clearly, before the rock mass is loaded, its initial hydraulic properties are isotropic, with $k_{x0}=k_{y0}=k_{z0}=1.30\times10^{-2}$ cm/s by Eq. (84). Under the condition of isotropic compression, the rock mass remains elastic, the isotropic property of hydraulic conductivity is maintained, and the magnitude of the hydraulic conductivity reduces by 2 orders of magnitude due to compression of fractures, with $k_x=k_y=k_z=4.82\times10^{-4}$ cm/s by Eq. (80). When shear stress is added incrementally on the rock mass block model from 0 to 20 MPa, the proposed method

As can be observed from Fig. 23, shear load has a substantial impact on the evolution of hydraulic conductivity of the rock mass model. Before the shear load reaches 4 MPa, the response of the rock mass model remains elastic, and the hydraulic conductivity components of the rock mass model are basically identical and do not vary with the shear load. When the shear load exceeds 4 MPa, however, hydraulic conductivity of the model becomes anisotropic. Due to shear dilation of fractures in the z-direction, the major hydraulic conductivities parallel to the direction of shear load in x-y plane, k_x and k_y, increase mildly at first when the shear load is smaller than 8 MPa. Afterwards, they increase dramatically, reaching an increase of 3-4 orders of magnitude. They approach a relatively stable state after the shear load increases up to 14 MPa. Obviously, the increase of k_x and k_y is resulted from the dilatancy behavior of the fractures related to equivalent plastic strain, as shown in Fig. 24, where the mobilized dilatancy angle approaches zero as the shear load approaches 14 MPa. When the shear load exceeds 14 MPa, shear dilatancy of the related fractures becomes trivial and hence k_x and k_y become steady. From Table 7 and Fig. 23, we can further see that k_x and k_y are very close to each other in values and they generally have the same varying trend with the increasing shear load.

Table 7. Major hydraulic conductivities of a cubic block of rock mass under isotropic compression and increasing shear loading

τ (MPa)	k_x (cm/s)	k_y (cm/s)	k_z (cm/s)	τ (MPa)	k_x (cm/s)	k_y (cm/s)	k_z (cm/s)
-	0.013016	0.013016	0.013016	10	0.279373	0.279350	0.000020
0	0.000482	0.000482	0.000482	11	1.088835	1.088816	0.000056
1	0.000482	0.000482	0.000482	12	2.204162	2.204158	0.000375
2	0.000482	0.000482	0.000482	13	3.171558	3.171559	0.001374
3	0.000483	0.000483	0.000482	14	3.676801	3.697449	0.022811
4	0.000494	0.000486	0.000474	15	3.915193	4.137786	0.224877
5	0.000543	0.000509	0.000444	16	4.063688	4.696511	0.635383
6	0.000657	0.000576	0.000372	17	4.243447	5.407600	1.167070
7	0.000742	0.000643	0.000282	18	4.635512	6.233203	1.600997
8	0.000704	0.000581	0.000207	19	5.390907	7.316177	1.928768
9	0.012562	0.012459	0.000106	20	6.462514	8.618240	2.159053

With the increase of shear load from 4 to 20 MPa, the change in the major hydraulic conductivity vertical to the direction of shear load, k_z, is even more interesting. Before the shear load reaches 10 MPa, k_z decreases significantly with increasing shear load and manifests a shear contraction-like behavior. When the shear load further increases, shear dilatancy occurs and k_z increases drastically, with changes in as high as 4-5 orders of magnitude. k_z reaches a relatively stable state after the shear load increases up to 17 MPa, which is actually a critical loading point that numerical instability may occur.

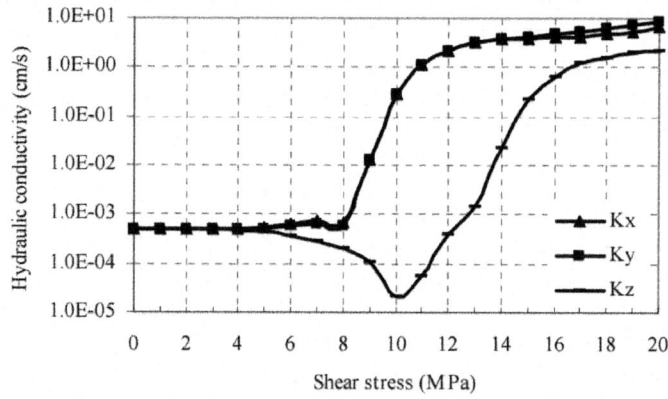

Figure 23. Major hydraulic conductivities of a cubic block of rock mass with increasing shear load.

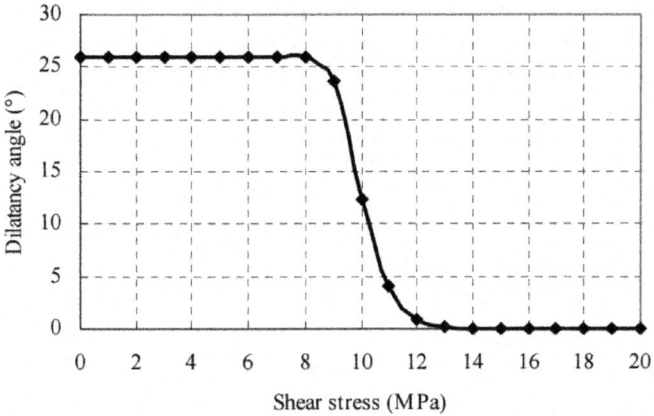

Figure 24. A typical case of mobilized dilatancy angle of a fracture with increasing shear load.

CONCLUSIONS

In this chapter, mathematical models were developed to estimate the hydraulic conductivity tensor for fractured rock masses subjected to mechanical loading or engineering disturbance. Emphases are placed on the investigation of the geological characteristics of rock masses as well as the coupling between fluid flow and stress/deformation, especially the effect of shear dilation or shear contraction on the hydraulic behavior of rock fractures.

The stress-dependent hydraulic conductivity tensor was formulated by using the superposition principle of flow dissipation energy on the basis of the concept of representative elementary volume (REV) and the assumption that rock masses can be treated as equivalent continuum media. The deformation behaviors of rock fractures subjected to normal and shear loadings are described with an elastic constitutive model, in which the pre-peak shear dilation or contraction of the fractures is empirically modelled. The validity of using the superposition principle of flow dissipation energy for development of the model is supported by the functional equivalence between the current formulation and the Snow's and Oda's models. This model is best suited for estimation of the hydraulic properties of rock masses at low stress level and with overall elastic response, and can be used to determine the applicability of the continuum approach to coupling analysis. The latter is achieved by performing numerical experiments to test the existence of the REV, and if exists, to further estimate the REV by gradually increasing the cubic volume of flow region, V_p, to see whether the hydraulic conductivity of the rock mass can eventually approach a steady point. The hydraulic properties and the REV size of the fractured rock mass at the construction site of the Laxiwa Hydropower Project were evaluated with the proposed model, and the calculation results were compared with the predictions of the Snow's model and validated by in-situ hydraulic tests, hence the feasibility of the proposed model in rock engineering practices is demonstrated.

The strain-dependent hydraulic conductivity tensor, on the other hand, was developed for disturbed rock masses under excavation or loading. In the model, a non-associative elasticperfectly plastic constitutive model was integrated to describe the deformation behaviors of the rock masses by characterizing them as equivalent continua containing one or multiple sets of parallel fractures. The clear advantages of the formulation are:

- The proposed hydraulic conductivity tensor is related to strains rather than stresses, hence enabling easier hydro-mechanical coupling analysis to include the effect of material nonlinearity of fractured rock masses.

- Beneficial from the equivalent non-associative elastic-perfectly plastic constitutive model, the hydraulic conductivity tensor considers the impact of shear dilatancy of fractures on fluid flow properties via mobilized dilatancy angles.

- When reduced to one dimensional case with a single fracture under normal and shear loadings, a closed-form solution to the hydraulic conductivity can be obtained, enabling validation of the model by laboratory coupled shear-flow tests of rock fractures.

- The proposed model is easy to be implemented in a FEM code, particularly suitable for numerical analysis of coupled hydro-mechanical processes in rock engineering.

The closed-form solution was validated by an existing coupled shear-flow test, and the evaluation results show that the proposed solution can closely describe the hydraulic behavior of a hard rock fracture under a wide range of normal and shear loads. The results of the simulation conducted to predict the excavation-induced hydraulic conductivities around a circular tunnel in a biaxial stress field at the Stripa mine are justified by in-situ experimental observations and compared with an existing elastic strain-dependent model, which show that engineering disturbance such as underground excavations may dramatically alter the hydraulic conductivities of the rock mass surrounding the excavations and change the isotropic pattern of the initial hydraulic conductivities. The numerical simulation on a cubic block model of a rock mass with three orthogonal sets of identical fractures under isotropic triaxial compression and shear loading further demonstrates that shear loading may drastically change the hydraulic properties of fractured rocks, in the magnitude of as high as 4-5 orders, and lead to high anisotropy of the hydraulic properties. Despite all these efforts, characterizing the hydraulic properties for fractured rock masses remains one of the most difficult research topics in rock mechanics. In the proposed models presented in this chapter, rock masses are assumed with rather regular distribution patterns of fractures, and the existence of a hydraulic conductivity tensor of the rock masses with any distribution of fractures is not discussed. The interaction between the fractures in the rock masses is also out of the scope of this chapter, and its effect on the hydraulic properties remains an open issue. Furthermore, the proposed models are established with a rather intuitive upscaling approach, and more rigorous homogenization schemes should be developed. All of there issues should be addressed in the future research.

ACKNOWLEDGEMENTS

The financial support from the National Natural Science Foundation of China (No. 51079107) and the National Natural Science Fund for Distinguished Young Scholars of China (No. 50725931), and the Program for New Century Excellent Talents in University (No. NCET-09-0610) for this study is gratefully acknowledged.

REFERENCES

1.   Alejano, L. R. & Alonso, E. (2005). Consideration of the dilatancy angle in rocks and rock masses. International Journal of Rock Mechanics and Mining Sciences, Vol. 42, No. 4, 481–507

2.   Barton, N. (1976). Rock mechanics review: the shear strength of rock and rock joints. International Journal of Rock Mechanics and Mining Sciences & Geomechanical abstracts, Vol. 13, No. 9, 255-279

3.   Barton, N. R. & Bandis, S. C. (1982). Effects of block size on the shear behavior of jointed rocks. In: Proc 23rd US Symp Rock Mechanics, Berkeley

4.   Barton, N.; Bandis, S. & Bakhtar, K. (1985). Strength, deformation and conductivity coupling of rock joints. International Journal of Rock Mechanics and Mining Sciences & Geomechanical abstracts, Vol. 22, No. 2, 121-140

5.   Bear, J. (1972). Dynamics of fluids in porous media. American Elsevier, New York

6.   Castillo, E. (1972). Mathematical model for two-dimensional percolation through fissured rock. In: Proc Int Symp Percolation through Fissured Rock, T1±D1–7, Stuttgart, Germany

7. Chen, Y. F.; Sheng, Y. Q. & Zhou, C. B. (2006). Strain-dependent permeability tensor for coupled M-H analysis of underground opening. Proceedings of the 4th Asian Rock Mechanics Symposium, pp. 271, Singapore, Nov 2006, World Scientific Publishing

8. Chen, Y. F.; Zhou, C. B. & Sheng, Y. Q. (2007). Formulation of strain-dependent hydraulic conductivity for fractured rock mass. International Journal of Rock Mechanics and Mining Sciences, Vol. 44, No. 7, 981-996

9. Chen, S. H. & Egger, P. (1999). Three dimensional elasto-viscoplastic finite element analysis of reinforced rock masses and its application. International Journal for Numerical and Analytical Methods in Geomechanics, Vol. 23, No. 1, 61–78

10. Chiarelli, A. S.; Shao, J. F. & Hoteit, N. (2003). Modeling of elastoplastic damage behavior of a claystone. Int J Plasticity, Vol. 19, 23–45

11. Esaki, T.; Du, S.; Mitani, Y.; Ikusada, K. & Jing, L. (1999). Development of a shear-flow test apparatus and determination of coupled properties for a single rock joint. International Journal of Rock Mechanics and Mining Sciences, Vol. 36, 641–50

12. Hoek, E.; Wood, D. & Shah, S. (1992). A modified Hoek-Brown criterion for jointed rock masses. In: Proc Rock Characterization Symp ISRM: Eurock 92, Hudson, J. A. (Ed.), 209–214, British Geotechnical Society, London

13. Hsieh, P. A. & Neuman, S. P. (1985). Field determination of the three-dimensional hydraulic conductivity tensor of anisotropic media. Water Resource Research, Vol. 21, No. 11, 1655–1665.

14. Huang, T. H.; Chang, C. S. & Chao, C. Y. (2002). Experimental and mathematical modeling for fracture of rock joint with regular asperities. Eng Fract Mech, Vol. 69, 1977– 1996

15. Indelman, P. & Dagan, G. (1993). Upscaling of permeability of anisotropic heterogeneous formations. Water Resources Research, Vol. 29, No. 4, 917-923

16. Jing, L. (2003). A review of techniques, advances and outstanding issues in numerical modeling for rock mechanics and rock engineering. International Journal of Rock Mechanics and Mining Sciences, Vol. 40, No., 283-353

17. Jing, L.; Stephansson, O. & Nordlund, E. (1993). Study of rock joints under cyclic loading conditions. Rock Mechanics and Rock Engineering, Vol. 26, No. 3, 215–32

18. Kelsall, P. C.; Case, J. B. & Chabannes, C. R. (1984). Evaluation of

excavation-induced changes in rock permeability. Int J Rock Mech Min Sci & Geomech Abstr, Vol. 21, No. 3, 123–35

19. Lai, T. Y. (2002). Multi-scale finite element modeling of strain localization in geomaterials with strong discontinuity. Ph.D. thesis, Stanford University

20. Liu, C. H.; Chen, C. X. & Fu, S. L. (2002). Testing study on seepage characteristics of single fracture with sand under shearing displacement. Chinese Journal of Rock Mechanics and Engineering, Vol. 21, No. 10, 1457-1461

21. Liu, J.; Elsworth, D. & Brady, B. H. (1999). Linking stress-dependent effective porosity and hydraulic conductivity fields to RMR. International Journal of Rock Mechanics and Mining Sciences, Vol. 36, 581-596

22. Liu, S. H. (1996). Generation of flow network and field tests on hydraulic conductivity for fractured rock mass. Northwestern Hydropower, Vol. 55, No. 1, 21-27

23. Lomize, G. M. (1951). Flow in fractured rocks. Gosenergoizdat, Moscow

24. Long, J. C. S.; Remer, J. S.; Wilson, C. R. & Witherspoon, P. A. (1982). Porous media equivalents for networks of discontinuous fractures. Water Resource Research, Vol. 18, No. 3, 645–58

25. Louis, C. (1971). A study of groundwater flow in jointed rock and its influence on the stability of rock masses. Rock Mechanics Research Report, No. 10, Imperial College of Science and Technology, London, Maini, YNT

26. Min, K. B. & Jing, L. (2003). Numerical determination of the equivalent elastic compliance tensor for fractured rock masses using the distinct element method. International Journal of Rock Mechanics and Mining Sciences, Vol. 40, No. 6, 795-816

27. Min, K. B.; Rutqvist, J.; Tsang, C. F. & Jing, L. (2004). Stress-dependent permeability of fractured rock masses: a numerical study. International Journal of Rock Mechanics and Mining Sciences, Vol. 41, No. 7, 1191-1210

28. Oda, M. (1985). Permeability tensor for discontinuous rock masses. Geotechnique, Vol. 35, No. 4, 483-195

29. Oda, M. (1986). An equivalent continuum model for coupled stress and fluid flow analysis in jointed rock masses. Water Resources Research, Vol. 22, No. 13, 1845-1856

30. Olsson, R. & Barton, N. (2001). An improved model for hydromechanical coupling during shearing of rock joints. International Journal of Rock

Mechanics and Mining Sciences, Vol. 38, No. 3, 317-329

31. Pande, G. N. & Xiong, W. (1982). An improved multilaminate model of jointed rock masses. In: Numerical Models in Geomechanics, Dungar, R.; Pande, G. N. & Studer, J. A. (Ed.), 218–226, Bulkema, Rotterdam

32. Patir, N. & Cheng, H. S. (1978). An average flow model for determining effects of threedimensional roughness on hydrodynamic lubrication. ASME Journal of Lubrication Technology, Vol. 100, 12-17

33. Plesha, M. E. (1987). Constitutive models for rock discontinuities with dilatancy and surface degradation. International Journal for Numerical and Analytical Methods in Geomechanics, Vol. 11, 345–62

34. Pusch, R. (1989). Alteration of the hydraulic conductivity of rock by tunnel excavation. Int J Rock Mech Min Sci & Geomech Abstr, Vol. 26, No. 1, 79–83

35. Snow, D. T. (1969). Anisotropic permeability of fractured media. Water Resources Research, Vol. 5, No. 6, 1273-1289

36. Vajdova, V. (2003). Failure mode, strain localization and permeability evolution in porous sedimentary rocks. Ph.D. thesis, Stony Brook University

37. Wang, M. & Kulatilake, P. H. S. W. (2002). Estimation of REV size and three dimensional hydraulic conductivity tensor for a fractured rock mass through a single well packer test and discrete fracture fluid flow modeling. International Journal of Rock Mechanics and Mining Sciences, Vol. 39, 887-904

38. Yuan, S. C. & Harrison, J. P. (2004). An empirical dilatancy index for the dilatant deformation of rock. International Journal of Rock Mechanics and Mining Sciences, Vol. 41, 679–86

39. Zimmerman, R. W.; Kumar, S. & Bodvarsson, G. S. (1991). Lubrication theory analysis of the permeability of rough-walled fractures. International Journal of Rock Mechanics and Mining Sciences, Vol. 28, No. 4, 325-331

40. Zhou, C. B.; Chen, Y. F. & Sheng, Y. Q. (2006). A generalized cubic law for rock joints considering post-peak mechanical effects. In: Proc GeoProc2006, 188–197, Nanjing, China

41. Zhou, C. B.; Sharma, R. S.; Chen Y. F. & Rong, G. (2008). Flow-Stress Coupled Permeability Tensor for Fractured Rock Masses. International Journal for Numerical and Analytical Methods in Geomechanics, Vol. 32, 1289-1309

42. Zhou, C. B. & Xiong, W. L. (1996). Permeability tensor for jointed rock masses in coupled seepage and stress fields. Chinese Journal of Rock

Mechanics and Engineering, Vol. 15, No. 4, 338-344

43. Zhou, C. B. & Xiong, W. L. (1997). Influence of geostatic stresses on permeability of jointed rock masses. Acta Seismologica Sinica, Vol. 10, No. 2, 193-204

44. Zhou, C. B.; Ye, Z. T. & Han, B. (1997). A study on configuration and hydraulic conductivity of rock joints. Advances in Water Science, Vol. 8, No. 3, 233-239

45. Zhou, C. B. & Yu, S. D. (1999). Representative elementary volume (REV): a fundamental problem for selecting the mechanical parameters of jointed rock mass. Chinese Journal of Engineering Geology, Vol. 7, No. 4, 332-336

Chapter 3

ROCK MASS GROUTING IN THE LØREN TUNNEL: CASE STUDY WITH THE MAIN FOCUS ON THE GROUTABILITY AND FEASIBILITY OF DRILL PARAMETER INTERPRETATION

Are Håvard Høien , Bjørn Nilsen

Norwegian Public Roads Administration, Postboks 8142 Dep, 0033 Oslo, Norway

ABSTRACT

The Løren road tunnel is a part of a major project at Ring road 3 in Oslo, Norway. The rock part of the tunnel is 915 m long and has two tubes with three lanes and breakdown lanes. Strict water ingress restriction was specified and continuous rock mass grouting was, therefore, carried out for the entire tunnel, which was excavated in folded Cambro-Silurian shales intruded by numerous dykes. This paper describes the rock mass grouting that was carried out for the Løren tunnel. Particular emphasis is placed on discussing grout consumption and the challenges that were encountered when passing under a distinct rock depression. Measurement while drilling (MWD) technology was used for this project, and, in this paper, the relationships between the drill parameter interpretation (DPI) factors water and fracturing are examined in relation to grout volumes. A lowering of the groundwater table was experienced during excavation under the rock depression, but the groundwater was nearly re-established after completion of the main construction work. A planned 80-m watertight concrete lining was not required to be built due to the excellent results from grouting in the rock depression area. A relationship was found between leakages mapped in the tunnel and the DPI water factor, indicating that water is actually present where the DPI water factor shows water in the rock. It is concluded that, for the Løren tunnel, careful planning and high-quality execution of the rock mass grouting made the measured water ingress meet the restrictions. For future projects, the DPI water factor may be used to give a better understanding of the material in which the rock mass grouting is performed and may also be used to reduce the time spent and volumes used when grouting.

INTRODUCTION

Tunnelling in urban areas involves many challenges. One of the biggest challenges is the risk of lowering the groundwater table, which may harm buildings and the environment. Urban tunnels usually have strict water ingress restrictions and systematic pre-grouting of the rock mass is usually described. In these cases, the rock mass grouting is a major part of the construction process and must be given great attention in the planning and execution phases in order to obtain a satisfactory result. In this paper, the typical challenges of such projects will be discussed based on the experiences from rock mass grouting in the Løren road tunnel, which is presently under construction in Oslo, Norway (see Fig. 1).

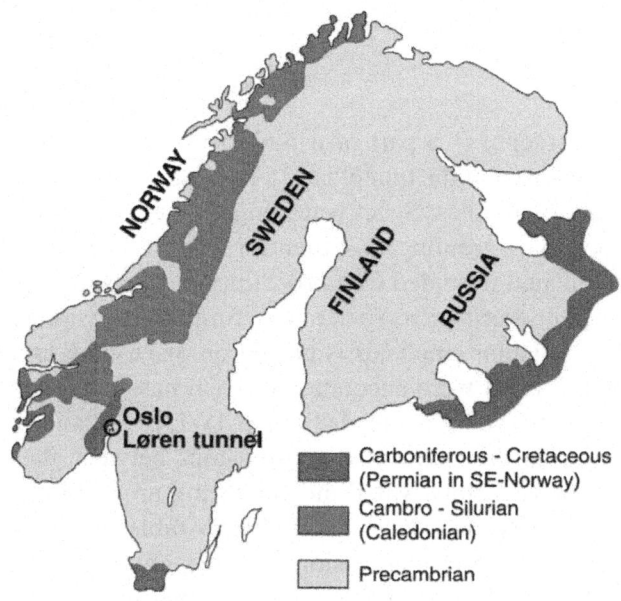

Figure. 1: Location of the Løren road tunnel, Oslo, Norway.

The rock part of the Løren tunnel is 915 m long (see Fig. 2). The tunnel has two tubes with three lanes and breakdown lanes, which results in face areas of 105–135 m². The tunnel is excavated from one side by drill and blast technology and continuous grouting. Tube B is excavated approximately

30 m ahead of tube A on descending profile numbers. In the area above the tunnel, there is a mix of residential and commercial buildings founded on soft, sensitive clay and sandy, gravelly soil, with a thickness of up to 30 m. Strict water ingress restrictions from 7 to 10 l/min/100 m were, therefore, given.

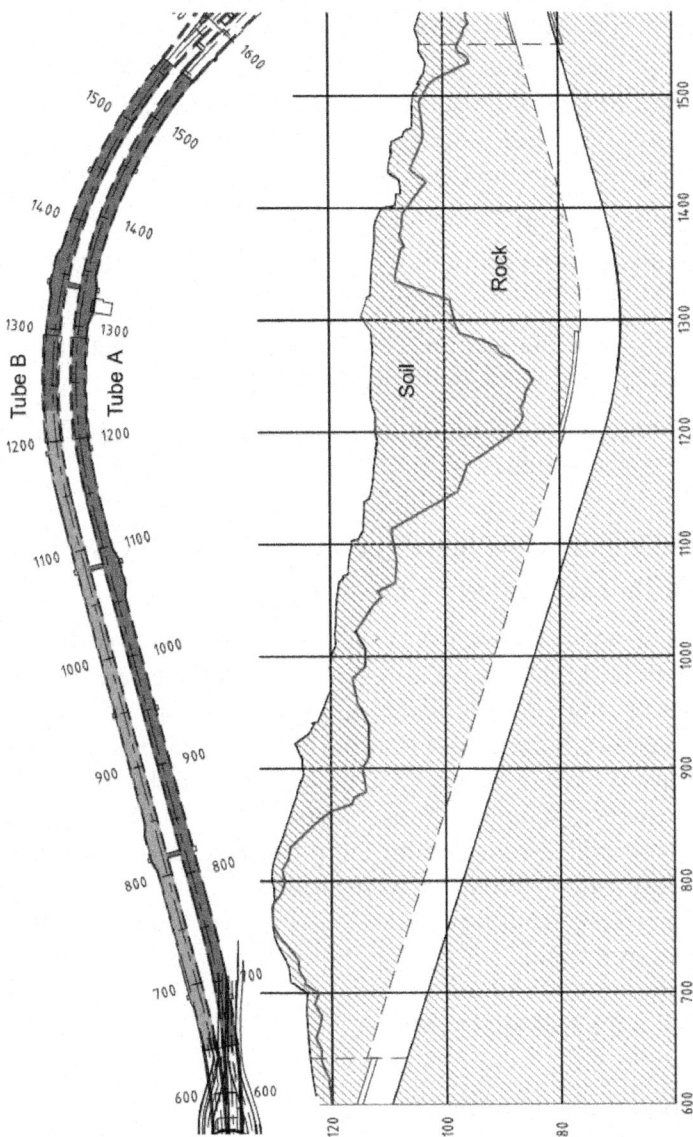

Figure. 2: Plan view and longitudinal profile of the Løren tunnel. The coloured areas on the plan view are the rock parts of the tunnel. The *orange area* is the section in focus on this article regarding drill parameter interpretation (DPI) and grout volumes.

This article will describe the rock mass grouting that has been done for the Løren project and present some issues and findings which are believed to be of particular interest. One such issue is groundwater and grouting aspects of the excavation under a depression with low rock overburden below 30 m of soil. Another issue which is discussed in more detail is the use of drill parameter interpretation (DPI) and its possible use for grouting purposes.

DPI is a new, innovative system for logging boreholes (measurement while drilling, MWD) and interpretation of these logs into factors that describe the rock mass. The DPI system at the Løren tunnel provided three factors: hardness, fracturing and water.

The main objectives in the following are, thus:

- To describe the geological and hydrogeological characteristics and the grouting procedures for the Løren tunnel.

- To describe the grouting results: in general, in relation to the rock depression and in relation to the grout consumptions.

- To describe the relationships between DPI water and fracturing factors and grout volumes

GROUND CONDITIONS AND GROUTING PROCEDURE FOR THE LØREN TUNNEL

The Løren tunnel is situated in a developed area with a mix of residential and commercial buildings. The residential buildings are mostly 1–2-floor townhouses or single-family houses constructed from wood and some larger concrete constructions with 4–7 floors. The commercial buildings are concrete constructions with 4–6 floors. An assessment of the potential settlement caused by lowering the groundwater level that could damage these buildings was, therefore, performed. During the pre-investigations, a large number of ground investigation boreholes were drilled to establish a theoretical rock surface and soil profiles. Observation wells were also established to survey the groundwater over time.

GROUND CONDITIONS

Soil and Rock Overburden

Throughout the tunnel, the rock overburden was mostly satisfactory, except for an area between profile nos. 1150–1300, where the rock surface has a depression as shown in Fig. 2. Due to alignment restrictions at the start and end point of the tunnel and road gradient restrictions, the rock overburden in this area was only 6–8 m for a 50-m stretch. The soil cover in the same area was approximately 30 m, with 3–5-m-thick dry crust clay on top of soft, sensitive clay, with sandy, gravelly soil at the bottom 8–10 m. This makes the area sensitive for settlement in case of lowering the groundwater level (Henriksen and Føyn 2004).

Groundwater

Before and during construction, 20 observation wells with automatic logging surveyed the groundwater level in the influential area for the tunnel. To obtain the seasonal and yearly variations in the groundwater levels, these wells had approximately 3 years of logging before the construction started. A few other wells had data from 5 years ahead of construction commencement. The long-term logging is important as a background when evaluating groundwater levels during construction.

Prior to the excavation, one infiltration well was established from the surface in the area mentioned above. In addition, in the same area, two infiltration holes were drilled during excavation from each of the tunnel tubes. These two holes were drilled along the tunnel axis with a small angle away from the tubes.

Bedrock Geology

During excavation, an engineering geologist mapped the geology, as illustrated in Fig. 3, and estimated a Qvalue after each blast. The geology information was registered in a tunnel documentation software called Novapoint Tunnel (Vianova 2011) on a fold-out tunnel profile (Humstad et al. 2012). A geological longitudinal section map as shown in Fig. 4 was made based on this mapping and the pre-construction investigations.

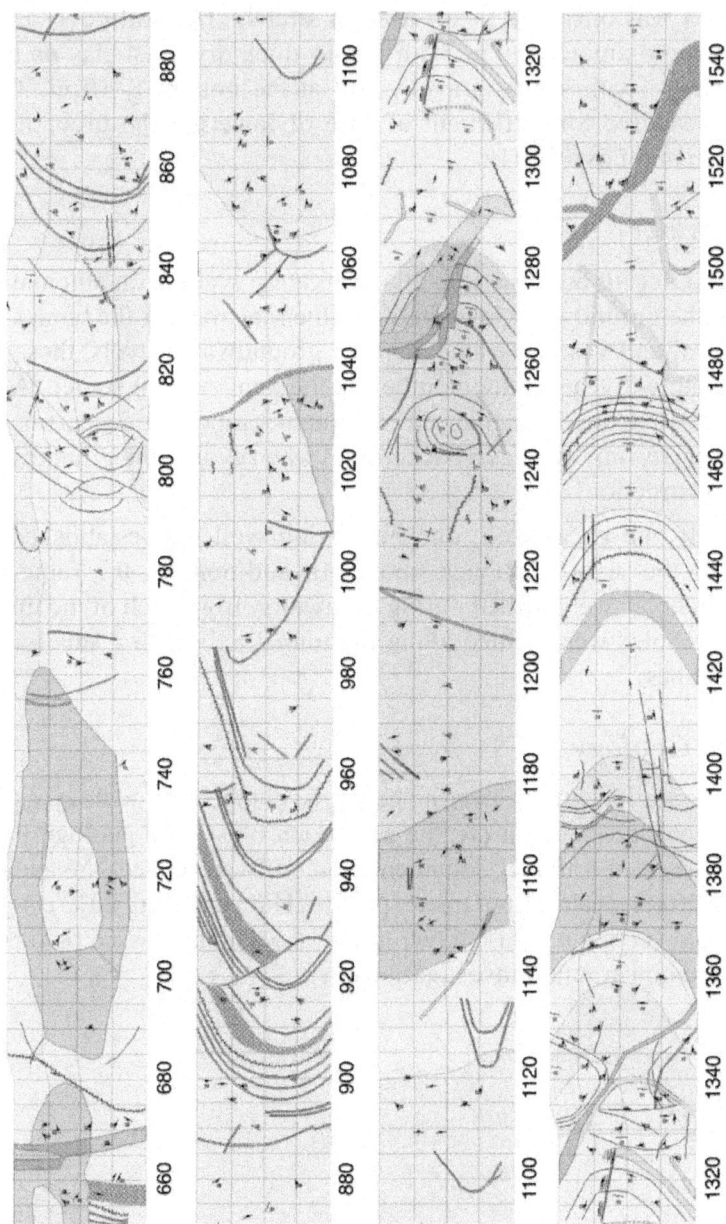

Figure. 3: Tunnel geology based on blast round mapping. The map is drawn as a fold-out tunnel profile. The rock types represented by the different colours are explained in Table 1. The *black lines* represent major joints and faults/weakness zones.

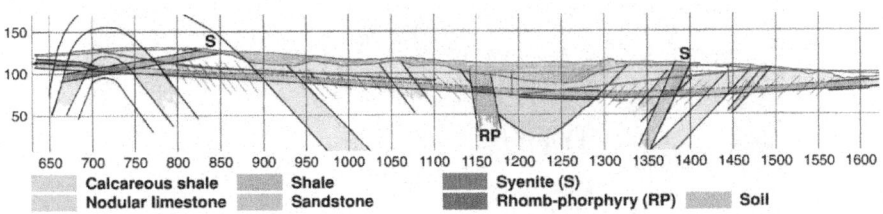

Figure. 4: Longitudinal profile illustrating the rock type distribution and structural geology (Iversen 2011).

The geology in the area, as shown in Fig. 1, is related to the Oslo Graben system of Cambrian, Ordovician and Silurian age. The rock types mapped in the tunnel are sandstones, calcareous shale, black shale, nodular limestone and intrusions of rhomb-porphyry, diabase and syenite. Details regarding the geology and rock type distribution along the tunnel are shown in Figs. 3, 4, 5 and 6, and Table 1.

Figure. 5: Typical geology of sedimentary rocks. Shale and bentonite rock at profile no. B952.

Figure. 6: Typical geology of intrusions. Sub-horizontal syenite intrusion and steep diabase dyke in sandstone (top) at profile no. A1239.

Table 1: Tunnel geology

From	To	Geology
B1560	B1445	Shale
B1445	B1425	Calcareous shale
B1325	B1420	Syenite
B1420	B1395	Nodular limestone
B1395	B1380	Black shale
B1380	B1355	Syenite
B1355	B1340	Nodular limestone
B1340	B1325	Shale, diabase
B1325	B1275	Alternating nodular limestone and black shale, diabase and syenite
B1275	B1175	Sandstone, diabase and syenite
B1175	B1145	Rhomb-porphyry
B1145	B1120	Nodular limestone
B1120	B1070	Shale
B1070	B1040	Nodular limestone
B1040	B980	Shale, with a small fault and syenite at B1040
B980	B885	Shale, with layers of bentonite rock
B885	B810	Shale
B810	B640	Calcareous shale with syenite in the tunnel roof

Weakness Zones and Fractures

The tunnel has intersected several minor faults/weakness zones, as shown in Fig. 3. Generally, the boundaries between sedimentary rocks and syenite/ rhomb-porphyry are weaker than the surrounding rock. Weakness zones from profile nos. 900–990 in Fig. 3 are layers of bentonite rock which crack/ disintegrate in contact with water.

The degree of fracturing along the tunnel, as mapped after each blast round, was mainly moderate (RQD mainly within the range 50–70), but with some variation, as illustrated in Fig. 7.

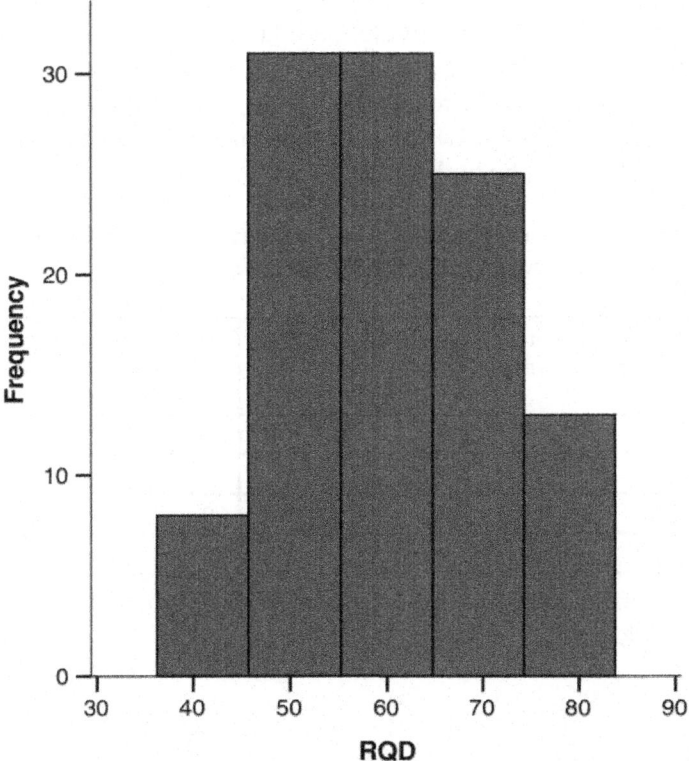

Figure. 7: Fracture frequencies (RQD) from *Q* value mapping during excavation.

Water Ingress Restrictions

During construction, thresholds across the tunnel floor were made to measure the water ingress, and water ingress restrictions (threshold values) were defined as shown in Table 2.

Table 2: Water ingress restrictions

From	To	Leakage for both tubes combined (l/min/100 m)
640	800	<10
800	1350	<7
1350	1500	<10
1500	1560	<7

Regarding the potential settlement of buildings, the depression between profile nos. 1300 and 1150 was the main concern during planning and excavation. Based on settlement calculations and empirical data of water ingress and pore pressure reduction in the Oslo region, a relationship was established between water ingress and settlement at profile no. 1250 (Henriksen and Føyn 2004), as shown in Table 3. This relationship was used during construction when deciding whether a planned watertight concrete lining between profile nos. 1220 and 1300 should be built or not.

Table 3: Relationship between pore pressure reduction, settlement and water ingress restrictions at profile no. 1250 (Henriksen and Føyn 2004)

Pore pressure reduction (m)	Settlement (cm)	Water ingress for both tubes combined (l/min/100 m)
1	3	3
3	8	7
5	12	10
10	20	20

An extensive monitoring programme was established to survey the potential settlement of buildings above the tunnel.

Grouting

Due to the very strict water ingress restrictions, maximum 7–10 l/min/100 m for both tubes combined, systematic pre-excavation grouting was planned and carried out for the entire tunnel.

Grout Rig

The grouting rig was equipped with three separate pumping lines and a storage cement mixer for each line. The cement was first mixed in the main cement mixer and then pumped into the storage mixers. There was also a tank and a pump for the accelerator to ensure controlled curing. The rig was operated by an electronic panel with predefined cement recipes.

Grout Curtains1

Grout curtains with 45–46 holes circumferencing the face and 7–9 holes in the face itself, as shown in Fig. 8, have been used. The hole length was 15, 21 or 23 m, depending on the distance between the curtains and the blast length, and the curtain overlap was 7–9 m. Spacings between the holes were 1.0 m at the floor and 0.8 m in the walls and crown.

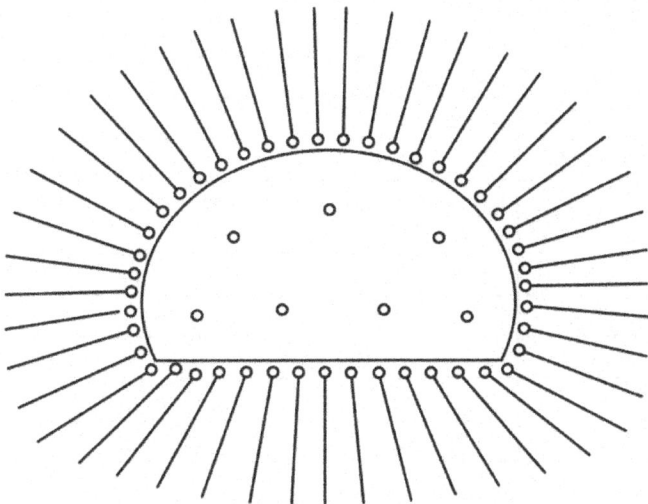

Figure. 8: Typical drill plan for grout curtains.

After completing the drilling of a curtain, grouting rods were placed in the holes with the tap open. For large leakages, the rods were placed in the holes during drilling with the tap closed. The grouting rods were 3 m long and had a disposable packer with a mechanical expansion sleeve at the end. The packer reflux valve was left open before grouting to let water pass, by inserting a nail in the cross opening.

Grouting Procedure

Cement has been used for all systematic grouting, normally with a w/c ratio span of 1.0 to 0.5, see Table 4. The grouting sequence started with two lines in the middle of the bottom holes. The lines were then moved from this position to both sides, and further up the walls, until they met at the crown. The third line was either used in the face holes or together with one of the other lines to grout two holes next to each other (see Fig. 9). The choice of grout cement type was mainly based on overburden evaluation and ingress restrictions. Rapid cement has the lowest cost and was, therefore, the preferred choice. Micro-cement is believed to have a better penetration than rapid cement (Axelsson et al. 2009) and was used when the overburden was low and high pressures might cause the grout to reach the surface and in cases when rapid cement did not give a satisfactorily tight tunnel. The cement was defined as rapid cement when it had a grain size $d_{95} > 20$ μm and micro-cement when it had a grain size $d_{95} < 20$ μm.

Table 4: Sections with different cement types

From	To	Cement type
B1550	B1301	Rapid cement
B1294	B1165	Micro-cement
B1159	B1022	Rapid cement, used some micro-cement in extra curtain at B1022
B1011	B0936	Micro-cement
B924	B762	Rapid cement
B757	B667	Micro-cement
B745		Single section with rapid cement
B1186		Single section with rapid cement

Figure. 9: Grouting in progress.

The stop criterion for grouting was when the pressure in the hole reached a certain value (the end pressure) and zero grout flow. In areas with small rock cover, the stop criterion (end pressure) was divided into bottom and top, where the top holes were defined as those above the invert (see Table 5).

Table 5: Grouting procedures for respective sections of the Løren tunnel

Date	Profile number	Hole length (m)	End pressure (bottom/top) (bar)	Description of the grouting procedure for a hole in the curtain
2009.06.23	B1550–B1535	23	45	Rapid cement
				340 l w/c 1.0
				345 l w/c 0.8
				330 l w/c 0.7
				600 l w/c 0.5
				At volumes <100 l, hold pressure at 80 bar for 5 min
				At volumes >1,500 l, the end pressure is 30 bar
2009.08.07	B1525	15	80	Same as above
2009.08.07	B1520–B1452	23	80	Same as above
2009.09.17	B1440–B1385	23	80	Rapid cement
				340 l w/c 1.0
				330 l w/c 0.7
				1,200 l w/c 0.5
				At volumes >1,870 l, controlled curing to stop grouting
				At volumes <100 l, hold pressure at 80 bar for 5 min
				At volumes >1,500 l, the end pressure is 30 bar
2009.10.22	B1379–B1346	18	80	Same as above

2009.11.15	B1333–B1301	18	80	Rapid cement
				340 l w/c 1.0
				330 l w/c 0.7
				1,200 l w/c 0.5
				At volumes >1,870 l, controlled curing to stop grouting
2009.12.08	B1294–B1274	19	80/60	Micro-cement
				340 l w/c 1.0
				340 l w/c 0.8
				800 l w/c 0.5
				At volumes >1,500 l, controlled curing if pressure <30 bar
2010.01.18	B1265	19	60/40	Micro-cement
				1,000 l w/c 0.8
				Controlled curing (2 %) at 800 l/hole
				Controlled curing (5 %) at 1,000 l to stop grouting
				Max. 9 m between grouting curtains
2010.02.02	B1258–B1239	15		Same as above
				Max. 6 m between grouting curtains
2010.02.22	B1230–B1178	15	60/40	Same as above
				Max. 8 m between grouting curtains
2010.04.08	B1174–B1165	15	60/40	Same as above
				Max. 4 m between grouting curtains

2010.04.23	B1161–B1154	15	80/80	Rapid cement
				340 l w/c 1.0
				330 l w/c 0.7
				530 l w/c 0.5
				300 l w/c 0.5 with controlled curing
				At volumes <100 l, hold pressure at 80 bar for 5 min
				At volumes >1,500 l, the end pressure is 30 bar
				Max. 9 m between grout curtains
2010.05.04	B1144–B1085	21	80/80	Rapid cement
				340 l w/c 1.0
				330 l w/c 0.7
				230 l w/c 0.5
				200 l w/c 0.5, controlled curing if pressure is <30 bar
				At volumes <100 l, hold pressure at 80 bar for 5 min
				Max. 10 m between grouting fans
2010.05.20	B1074–B1122	23	80/80	Same as above
				Max. 12 m between grout curtains
2010.08.11	B1011–B935	23	60/40	Micro-cement
				800 l w/c 0.8
				200 l w/c 0.8, controlled curing 2 %
				At volumes >1,000 l, controlled curing 5 %
				Max. 12 m between grouting curtains

2010.09.21	B924[a]– B901	23	60/40	Rapid cement
				700 l w/c 0.7
				500 l w/c 0.5
				At volumes >1,200, grouting is ended
				Max. 12 m between grouting curtains
2010.10.06	B892– B762	23	60/60	Rapid cement
				300 l w/c 1.0
				300 l w/c 0.7
				400 l w/c 0.5
				After 900 l and pressure <30 bar, a break in grouting is taken
				Max. 12 m between grouting curtains
2010.12.03	B757[b]– B735	23	60/60	Micro-cement
				300 l w/c 0.8
				300 l w/c 0.5
				150 l w/c 0.5, controlled curing 3 %
				150 l w/c 0.5, controlled curing 6–8 %
				Max. 12 m between grouting curtains[c]
2011.01.03	B723– B767	23	60/40	Same as above

[a]The grouting procedure used from B924 was changed because the grout volumes were too low to obtain a tight tunnel

[b]There was grouting twice at B762, first with rapid cement and then micro-cement, because of an initially poor grouting result. Only the rapid cement grouting curtain is included in the dataset

[c]On some occasions, the distance is 6 m

When grouting, it is generally important to have some grout consumption and pressure build-up in the hole in order to make a tight tunnel (Brantberger et al. 2000). Therefore, grouting at the Løren tunnel, as shown in Table 5, was started with a highly liquid grout, which was made less liquid by lowering the w/c ratio (for rapid cement) or adding curing liquid (for micro-cement) to build up pressure during the grouting process for the hole. If there was no sign of pressure build-up after this, the hole was set on hold (e.g. for an hour) before grouting was resumed.

For the controlled curing, the accelerator was mixed with the grout at the rod. The contractually described controlled curing for rapid cement (see Table 5) did not work due to the chemical composition of the cement.

During construction, it was detected that the pumping system for the accelerator was not operating as it was supposed to. It was found that the pump fed a lower amount than the percentage shown on the panel. Towards the end of the construction time (around profile no. B757), this was taken into account by using a higher setting than that described.

MWD AND DPI

Background and Basic Principles

Measurement while drilling (MWD has a history from the early 1900s when Schlumberger introduced downhole electrical logging to the oil industry (Segui and Higgins 2002; Smith 2002). When the mining industry started to adapt the method, the logged data were plotted on paper with pen strip recorders and the drill parameters were inspected manually. When computers were introduced, new possibilities with filtering and advanced statistical analysis became available (Schunnesson 1997) for proper drill parameter interpretation (DPI). The latest software on the market uses 3D technology to present the data and, together with easy-to-use interfaces, the method is now a good tool to predict the rock conditions for mining and tunnelling purposes. During drilling, the parameters from the machine control are logged and saved in files. These data are stored together with the hole position, hole direction, drill hammer ID and other relevant data. An example of the logged parameters is shown in Table 6.

Table 6: Example of logged parameters from a log file

HD (mm)[a]	PR (dm/ min)[b]	HP (bar)[c]	FP (bar)[d]	DP (bar)[e]	RS (r/ min)[f]	RP (bar)[g]	WF (l/ min)[h]	WP (bar)[i]	Time
32	0	69.76	19.69	39.65	216.86	42.47	76.09	21.4	8:18:49
59	31.83	101.44	17.12	40.08	217.77	38.61	76.84	20.97	8:18:49
84	30.28	111.71	17.12	40.08	218.68	39.47	75.72	20.97	8:18:50
108	28.37	116.42	19.26	40.51	219.60	42.04	75.72	21.83	8:18:50
131	26.75	123.26	20.97	40.51	214.11	53.20	76.46	20.97	8:18:51
152	24.88	124.98	24.82	40.51	198.56	51.48	76.46	22.26	8:18:52

[a]Hole depth

[b]Percussion rate

[c]Hammer pressure

[d]Feeding pressure

[e]Dampening pressure

[f]Rotation speed

[g]Rotation pressure

[h]Water flow

[i]Water pressure

In tunnelling and mining, the DPI usually gives three factors that describe the rock condition: rock hardness, fracturing and water. The levels of these three factors can be combined to describe other types of rock properties, e.g. drillability and blastability, and it can be evaluated, for instance, when to start spiling and rock mass grouting.

In the case of mining for example, the hardness factor can be used to find the boundary between the ore body and the side rock. In tunnelling, a common use of DPI is to evaluate the quality of the rock mass in exploratory boreholes in order to adjust the support level when advancing towards weakness zones.

The water and fracturing factors are not yet very commonly used, but one example is Norra Länken in Stockholm, where the interpreted parameters were evaluated while drilling the grout curtain. When starting on a new curtain, every second of the holes were drilled first and the drill logs sent to the site office. An engineering geologist evaluated the interpreted data and ordered extra holes if required according to the DPI grouting class (Carlsvärd and Wallgren 2009).

The DPI software that has been used at the Løren tunnel is GPM+ Tunnel, delivered by Rockma AB (2011). GPM+ calculates three factors: hardness, fracturing and water. The values are on a relative scale, but the hardness factor can be adjusted based on rock strength tests (e.g. Schmidt hammer or point load) to estimate the uniaxial strength (Valli 2010).

GPM+ uses different statistical methods and filtering for the calibration and processing of data. The calibration must be done at each work site and for each drill rig in order to obtain a satisfactory result. The calibration is done based on a principal component analysis to discover which of the parameters are relevant, and to what degree they are relevant. The algorithms are based on the work of Schunnesson (1997). For calculating the DPI factors, GPM+ uses the values from the parameters in the log file (see Table 6) for each sample, and returns a value for each of the DPI factors (hardness, fracture and water) for each sample point. The hardness is, in this case, calibrated to show the uniaxial strength. The fracturing factor indicates that the rock is fractured and the water factor indicates that water is flowing from the rock. A higher value indicates a higher degree of fractures or more water. In GPM+, the water factor is set to show water for values higher than 1.5 and to show fracturing for fracturing values higher than 1.0. See Figs. 10 and 11.

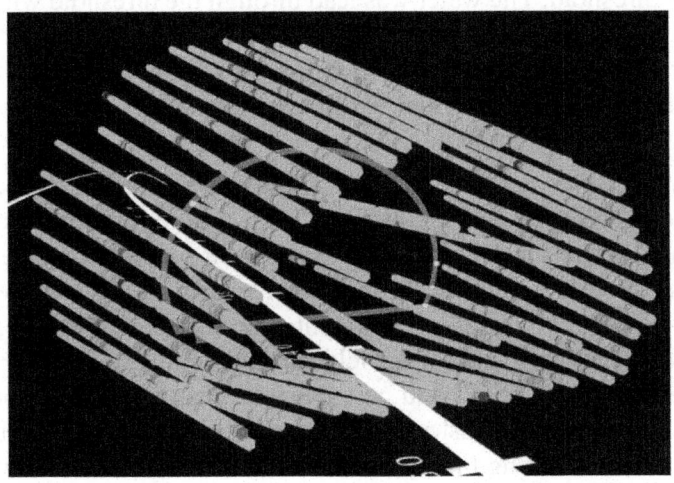

Figure. 10: Screenshot from GPM+ showing the water factor for a grouting section in 3D.

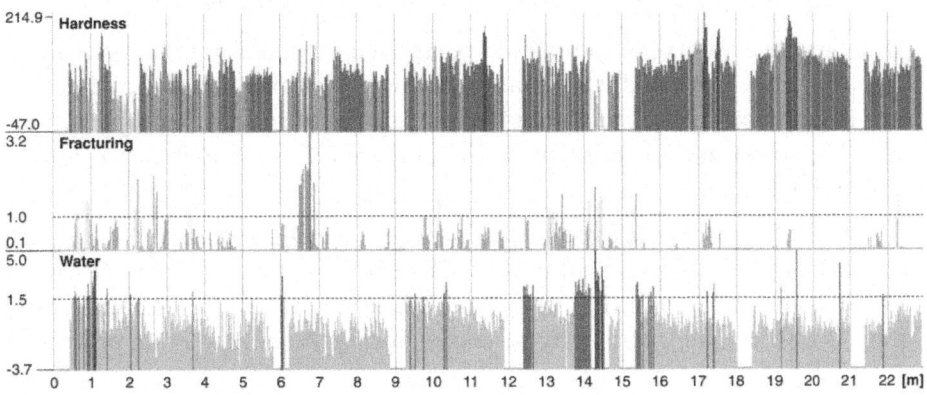

Figure. 11: Screenshot from GPM+ showing (from the top) hardness, fracturing and water along one borehole

Data Collection

Water Ingress Measurement

As mentioned in Sect. 2.2, thresholds across the tunnel floor were made to measure the water ingress during construction. The construction road runs over the thresholds, with a ditch on the side for the water to flow and stop behind the threshold. The water was lead through the threshold with a pipe and the water volume was measured by collecting the water in buckets for 1 min. The water volume from the upstream threshold was subtracted and the spacing between them taken into account by dividing the distance and multiplying by 100 (m) to obtain l/min/100 m.

The main concern by using this method is that blasting and crushing by large vehicles creates mud and silt that fills up behind the thresholds and clogs the drainage in the top layer of the construction road. The space behind the thresholds, therefore, have to be maintained regularly by removing the silt and clay. To lead all the water into the ditch and, at the same time, having the construction road driveable for vehicles is challenging, and water may easily follow the construction road and pass over the thresholds. Also, other uses of the tunnel space during construction, e.g. storing and temporary installations, may conflict with leading the water into the thresholds. In addition, an area at the low point lacks water ingress measurement because of the cross-cut and space for water sedimentation.

Leakage Mapping and Comparison with the DPI Water Factor

After the excavation of the tunnel was completed, a manual mapping of leakages from the tunnel roof and walls was performed. The mapping covered point as well as area leakages. The volume of leakages is measured in drops/min or l/min. The registration is done on a fold-out tunnel profile and is registered in Novapoint Tunnel.

In GPM+, there is an export function where the DPI factors are interpolated and saved to a fold-out tunnel profile. These files were imported to Novapoint Tunnel and can be viewed together with the water leakage information. Small impacts of the water factor may be erased during the interpolation process.

Processing of DPI and Grouting Data

To investigate the relationship between DPI and grouting data, some processing of the DPI data and matching of DPI and grouting data had to be done. The dataset consists of data from two sources: interpreted drill data done by GPM+ Tunnel and grout logs from the grouting rig. The statistical software program

SPSS (IBM 2010) was used to pair up these datasets and recalculate DPI values. The program has also been used for descriptive purposes and chart building.

Interpreted Drill Data

GPM+ mainly shows the interpreted data graphically, with color scales representing different values. For the purpose of statistical analyses and charts, number values are needed, and Rockma Systems AB, therefore, exported all measuring points (one point every 2–3 cm in all holes) into text files. These text files were imported into SPSS and resulted in a dataset with over 2 million measuring points.

To compare the interpreted data with the grouting data, one value for each factor (water and fracturing) for each hole is needed. Based on the experience of Rockma Systems AB, values for water higher than 1.5 indicate water-bearing fractures and fracturing values higher than 1.0 indicate fractured rock. To provide one value for each hole, a filtering and recalculation for each factor was performed.

For the water factor, all sample values ≤1.5 were filtered out. Sample values ≥1.5 were subtracted 1.5 and multiplied with their sample length and then added:

$$W = \sum_{i=m}^{n}(((w \geq 1.5) - 1.5) * l)_i \tag{1}$$

where W is the water value for a hole, n the number of samples for a hole, w a single water value and l is the sample length.

For the fracturing factor, all sample values ≤1.0 were filtered out. Sample values ≥1.0 were subtracted 1.0 and multiplied with their sample length and then added:

$$F = \sum_{i=m}^{n}(((f \geq 1.0) - 1.0) * l)_i \tag{2}$$

where F is the fracturing value for a hole, n the number of samples for a hole, f a single fracturing value and l is the sample length.

W and F factors were then added in a new dataset, along with their section numbers, hole numbers and hole lengths. This dataset contained 3,611 cases (holes).

When water and fracturing values are presented for a section, the W and F factors are added for that section. In the results for the water factor, it is a trend that the impact is larger with decreasing profile number.

A linear trend line was calculated and the difference quotient for this line has been used to normalize the data by recalculating the values to have a horizontal trend line, to make them comparative through the tunnel.

Grouting Data

The grouting data were obtained based on the log files from the grouting rig and saved as spreadsheets containing the weight of cement used and the amount of grout in liters for each recipe and hole. The values for each hole were added, giving the weight of cement and liters of grout for each hole. The weight of cement was divided into three categories: rapid cement, micro-cement and micro-cement with controlled curing.

Hole Number Matching

The hole numbers from grouting data and drill data were not corresponding. The hole numbering for the grouting data started at the lower right corner and proceeded clockwise around the contour, with the face holes as the last numbers. The hole numbering for the drill data was more or less random, but could be viewed together with the drill pattern in GPM+.

The hole numbers, therefore, were punched manually to make the two data sources correspond. For the start of the tunnel, it was hard to decide where to start the hole numbering, and this part is, therefore, excluded from the dataset, as well as a couple of sections where there was doubt about where the hole numbering started. All face holes are also excluded from the dataset due to difficulties with the hole numbers matching.

Content of the Dataset

The dataset spans over profile nos. B1224 to B641 and has 59 different sections/grouting rounds and 2,606 cases/holes. This data was picked out because tube B was excavated approximately 30 m ahead of tube A and the grouting, therefore, is done without any disturbance from the other face. Hole number matching was difficult between B1560 and B1224, and this section was, therefore, excluded.

The main variables in the dataset are:

- Section number

- Hole number

- Hole length

- Hole number from drill data

- Hole number from grouting data

- Rapid cement weight

- Micro-cement weight

- Micro-cement with controlled curing weight

- Grout volume

- W: water factor for a hole

- F: fracturing factor for a hole

RESULTS FROM THE TUNNELLING

The presentation of the results is divided into two main parts, grouting and DPI. The grouting part has its main focus on describing water ingress and on how the challenging grouting past the rock depression between profile nos. 1300 and 1150 was handled. Some findings regarding grout consumption in holes are also presented. The DPI part focuses on a comparison between the grouting volumes and DPI factors for selected sections.

Grouting

Water Ingress

As shown in Table 7, the water ingress values are generally below the restriction (threshold) values for one tube, and sometimes slightly exceed the restrictions when the values for the two tubes are added.

Table 7: Water ingress for tubes A and B measured 03.01.2011

From	To	l/m	l/m/100 m
A1545	A1355	6.0[a]	3.2[a]
A1300	A1220	2.7	3.4
A1220	A1050	8.0	4.7
A1050	A749	4.0	1.3
B1562	B1385	12.0[b]	6.8[b]
B1303	B1232	3.0	4.2
B1232	B1060	5	2.9
B1060	B723	7.6[c]	2.3[c]

[a]Measured 05.04.2012

[b]Measured 03.01.2012

[c]Measured 02.01.2011. There are some uncertainties regarding this measurement

The sampled distances do not match the ingress restrictions because the placing of the thresholds had to be adjusted to the placing of low points, cross-cuts etc. Because of this, some assessment of the values had to be made when considering whether the restrictions are met or not.

The 7 l/min/100 m restriction for profile nos. 1560–1500 is a part of a bigger sample interval, but when looking at the leakage mapping, there are not many leakages in this area and, therefore, it is probably below the restriction value. The same assumption applies for profile nos. 800–640, where a part of this interval is not a part of a sample.

Passing the Rock Depression, Profile nos. 1300–1150

The biggest challenge during the construction of the Løren tunnel was passing the rock depression between profile nos. 1300–1150. As mentioned before, the rock overburden was 6–8 m and the soil thickness was 30 m for a 50-m section in this area. The excavation through this section was done with 3-m blast rounds, and spiling, shotcrete, rock bolts and rebar-reinforced shotcrete ribs as rock support. Exploratory boreholes from the tunnel face were drilled to make sure that the overburden was as expected.

Because of the potential consequences for buildings at the surface if the water ingress was too high after rock mass grouting, it was prepared for a watertight concrete lining between profile nos. 1300–1220. During excavation, a larger profile with an arched floor was blasted to make room for this construction.

Due to problems during rock mass grouting around the building pit, the water level in the depression was lowered 5–6 m before tunnel excavation started. This lowering was expected to be temporary because barriers and watertight constructions would allow the water level to rise in the building pit when excavation was finished.

To raise the groundwater level and reduce the settlement, infiltration from the surface was started prior to tunnel excavation. In addition, infiltration from the two infiltration boreholes in the tunnel tubes was started when the excavation had passed the depression.

To decide whether the watertight lining should be built or not, a thorough evaluation of water ingress, groundwater level and ground reaction due to the lowering of the groundwater had to be made. To get a better understanding on how the infiltration affected the groundwater level, the infiltration was stopped for two periods. The curves from the observation wells were flattening out after approximately 14 days and showed a lowering of the groundwater level by approximately 2 m (Boge 2010).

The calculated settling was 50–100 mm, and without water infiltration, an additional 50 mm. The registered settlements of buildings were smaller than expected and it is reason to believe that soil is more pre-consolidated than the laboratory testing shows. Levelling measurements on a building founded on rock showed a 30-mm lifting, which is assumed to be a result from hydraulic splitting of the rock from the rock mass grouting (Boge 2010). Measurements done on buildings on other parts of the tunnel also indicate this.

When deciding whether the watertight lining should be built or not, a conservative assumption was made of further settling of 70–100 mm if the water infiltration was turned off. This was believed to represent minor harm for buildings. It was also concluded that the building of the watertight lining would have a minor influence on the groundwater level (Boge 2010). It was, therefore, decided not to build the lining and to turn off the water infiltration (Barstad 2010).

The grouting in the rock depression area resulted in water ingress corresponding to the requirements in the contract. After the groundwater level in the building pit was re-established, the groundwater level in the area is nearly the same as before the project started. Settlements of the most exposed buildings are in the order of magnitude of 30–40 mm. The settlement seems to have stopped and there has been little damage to buildings due to groundwater lowering (Boge 2012).

Grout Consumption

The grout volume frequencies for all holes are shown in Fig. 12. The histogram illustrates that a large amount of holes had a grout consumption less than 100 l. For most of the grouting procedures in the dataset, the maximum grout volume for a hole was 1,000 l. For holes with a consumption of more than about 1,000 l, the end pressure was hard to achieve. Holes with a grout consumption above 1,250–1,500 may be a result of flow back into the tunnel through fractures.

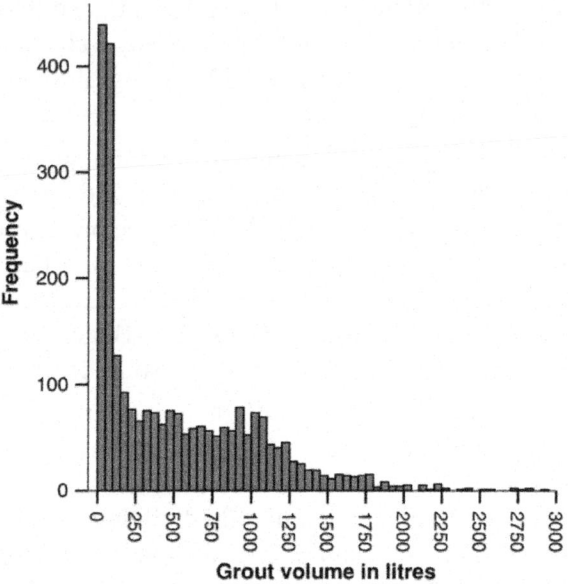

Figure. 12: Grout volume frequencies for all holes, $N = 2,606$.

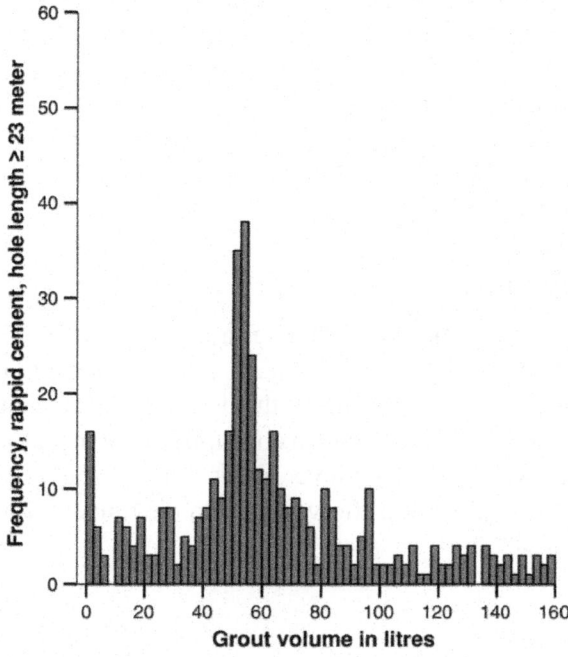

Figure. 13: Grout volume frequencies of holes filled with rapid cement and hole length ≥ 23 m (grout volume is cut at 160 l), $N = 986$.

In Fig. 13, a detailed frequency diagram is shown for holes with 23 m length grouted with rapid cement. The theoretical volume of these holes is 47 1,2 but the volume might be larger because the actual hole diameter might have been wider and because fall-out of small rock pieces might have occurred. In addition, pressure buildup in the hoses and filling of the rods uses some grout. The peak around 52 1 is, therefore, an indication that a large part of the holes only is filled with grout, with no penetration into the rock mass.

As shown in Fig. 14, holes grouted with micro-cement does not have a peak at around 55 1, as in the holes grouted with rapid cement. A large amount of holes with almost no grout consumption can be observed in this figure, in contrast to what can be observed for the rapid cement holes in Fig. 13.

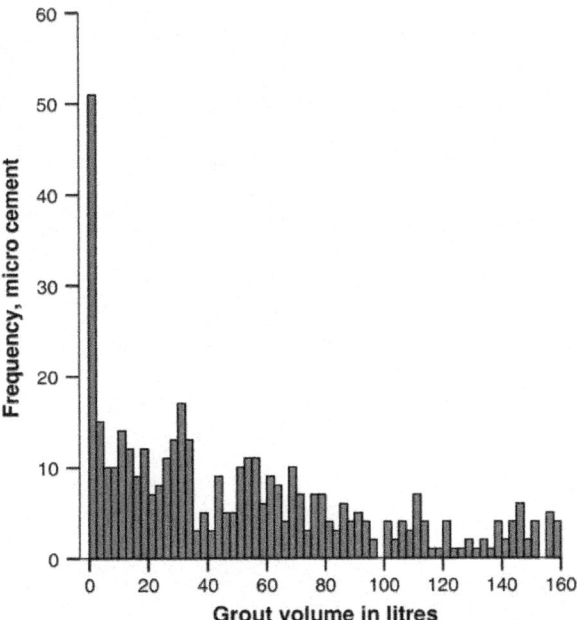

Figure. 14: Grout volume frequencies of holes grouted with micro-cement (grout volume is cut at 160 1), $N = 1,098$.

Drill Parameter Interpretation

Water Leakage

The mapped water leakage has a generally good match with the DPI water factor, although the mapping is done after the grouting was finished. This match indicates that, when the water factor shows water, water is quite likely actually present in the hole. Since the water may find new channels during grouting and positioning while mapping is challenging, an exact match cannot be expected. An example of mapped leakage versus water factor from DPI is shown in Fig. 15.

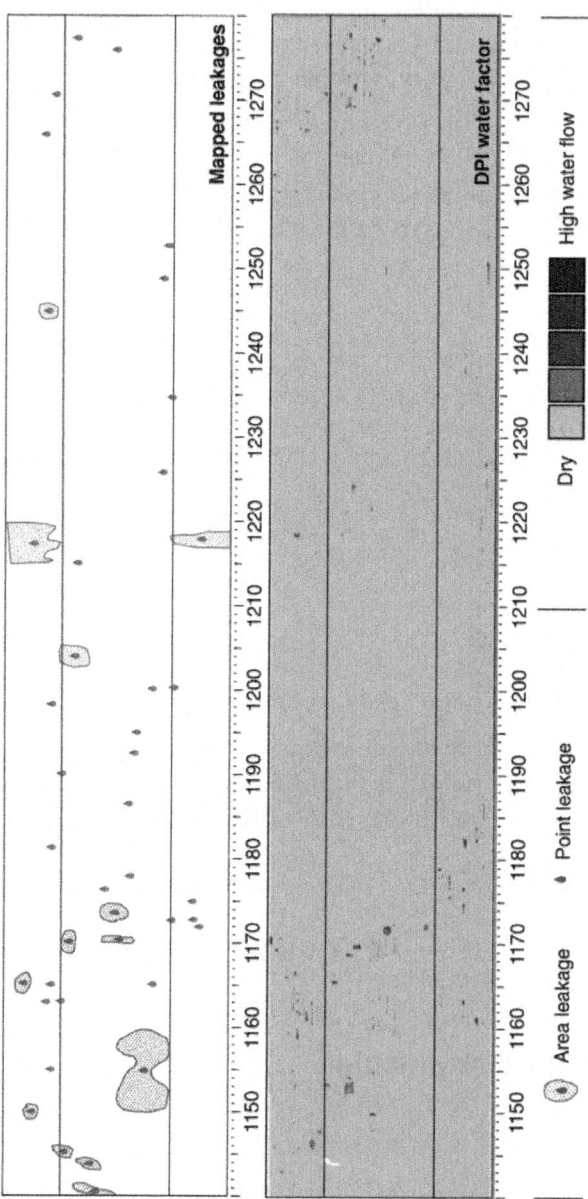

Figure. 15: Leakage mapping in the tunnel (*upper part*) and interpolated water factor from DPI (*lower part*).

Grouting Volumes and the DPI Factors Water and Fracture

The water factors for the respective sections are plotted in Fig. 16. A trend indicating that the water factor is increasing (seen in the excavation direction) from B1224 to the end of the tunnel (plotted with a red linear line) can be observed. Based on this trend line, the water factor has been normalised as described in Sect.3.3.

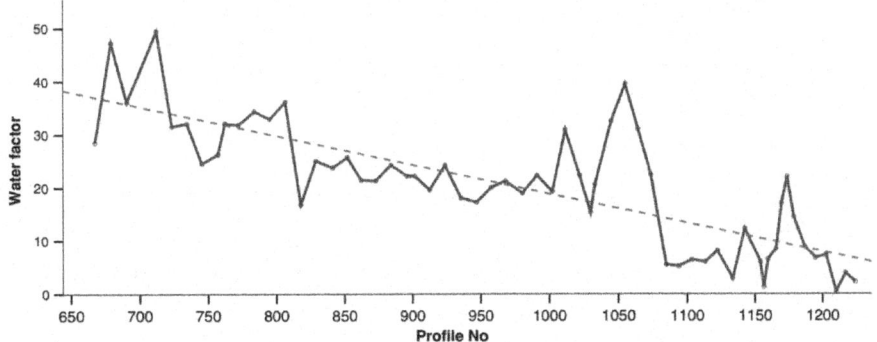

Figure. 16: Water factor for each section, along with the trend line used in the normalization.

In Figs. 17 and 18, the normalised water factor and the fracturing factor, respectively, and the grout volumes are plotted versus the profile number, and the main grouting parameters are given. The curves are simply fitted over each other to display possible connections between the parameters.

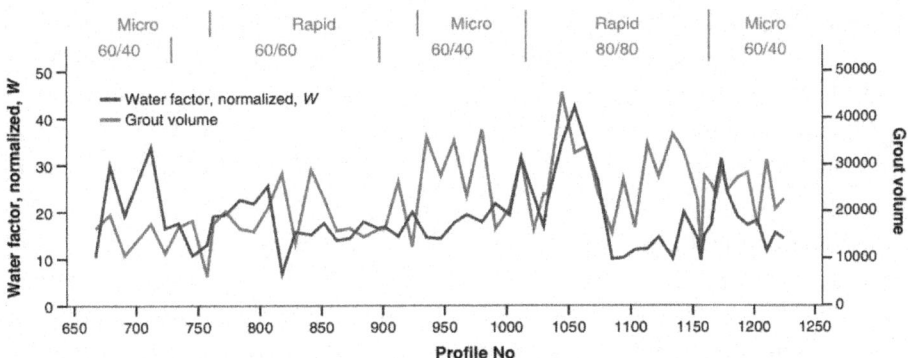

Figure. 17: Normalized water factor and grout volume for each section (grouting parameters at the top).

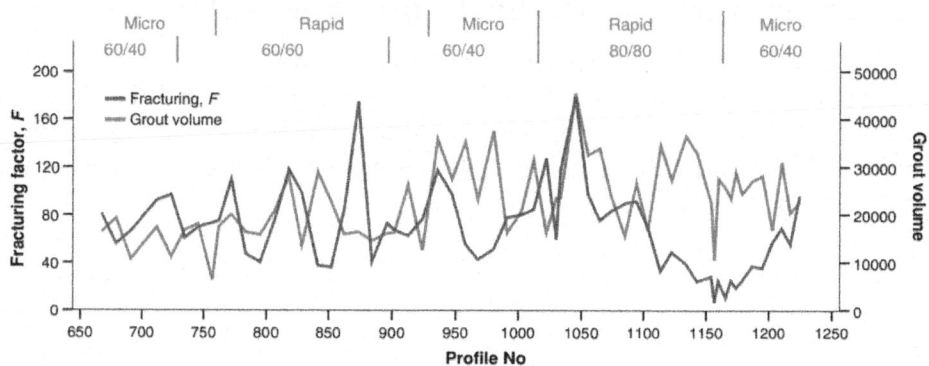

Figure. 18: Fracturing factor and grout volume for each section (grouting parameters at the top).

For Fig. 17, some sections seem to have fairly good correlation between the water factor and grout volumes, i.e. for B1203–B1154 and B1085–B991, the lines have a relatively good match. For B735–B667, the lines follow each other's paths, but have a scale difference. Also, an area with less fluctuation, B901–B862, may have a certain relationship.

In Fig. 18, there are a few small areas that might indicate a relationship between the fracturing factor and grout volumes, i.e. B1055–B1033 and B784–B762.

DISCUSSION AND CONCLUSIONS

The pre-injection grouting that has been carried out for the Løren tunnel has given a water ingress which is lower than the pre-defined very strict restrictions, and, therefore, seems to have been successful. The reservation, however, has to be taken that this tunnel project is not quite completed, and there is a chance that the water ingress may increase when the groundwater level is back to normal.

As shown in Table 5, the grouting procedures were continuously modified during the construction process. To obtain a satisfactory grouting result, it is important to always evaluate the grouting procedures and observe the water leakages in the tunnel. At the Løren tunnel, the grouting procedures were adjusted to ensure satisfactory grout consumption and the possibility for the pressure to build up. Another adjustment for ensuring a satisfactory grouting result was the regulation of distances between curtains and grout hole lengths.

The experience from the Løren tunnel also emphasises that, during tunnelling, it is very important to observe the groundwater level and settlement of buildings for potential reconsideration of settlement calculations, need for extra waterproofing and/or need for permanent water injection. In the Løren case, based on such considerations, a planned watertight concrete lining and water infiltration system were found not to be required. The temporary infiltration boreholes in the tunnel were made permanent in case settlement on buildings should resume (but are not in use).

During the grouting operation, a large portion of the holes grouted with rapid cement was filled with approximately 52 l of grout, while for the holes grouted with micro-cement, this same trend was not found (as shown in Figs. 14 and 15). This indicates that only the holes grouted with rapid cement are filled, and that there is no penetration into the rock mass. This illustrates that, in areas where water is expected, it should always be considered to switch to micro-cement in order to penetrate narrow water-bearing fractures.

It was also observed that a large portion of the holes grouted with micro-cement had almost no grout consumption (see Fig. 14). This may have been caused by micro-cement intrusion from nearby holes, with controlled curing, which cause the cement from nearby holes to harden and impede the grouting. Because of this, one should limit the quantities used when grouting with controlled curing. Another large portion of the holes in Fig. 14 range from very small grout consumption to about 60 l. A possible cause of the low grout consumption in these cases may be explained by micro-cement without controlled curing which may have intruded from distant holes, partially hardened and interfered with the grout consumption.

As explained in Sect. 2.3, it was detected during construction that the cement accelerator pump fed a lower amount of grout than it was supposed to while performing controlled curing. The pump used was a rotation pump, which was not suitable for this application. It should, therefore, be emphasised that a piston pump, which can follow the strokes and flow of the cement pump, should be used when dealing with accelerator. This will ensure the right amount of accelerator for a successful controlled curing process.

The DPI and grout volumes in the Løren tunnel indicate that there may be a relationship between the grout volume and water factor if special geological features are taken into account.

There are many factors that may have had an influence on the grouting volumes and DPI factors. The main variables that may affect the results are as follows:

- Structural geology
- Rock properties

- Rock cover

- Rock stress

- Surrounding terrain/topography

- Type of grout and w/c

- End pressure

- Groundwater level

- The order in which the holes are drilled

- The order in which the holes are grouted

From approximately B990–B900, layers of bentonite rock have probably affected the grouting volumes. During grouting in this area, grout reached the surface on several occasions. The area above the tunnel had a lifting of the ground likely caused by the grouting and grout was observed as layers in the bedding plane direction during mapping. This indicates that hydraulic fracturing of the rock has caused new fractures that were not registered by the DPI. The peaks in the water parameter seen in Fig. 14 at B1173 and B1055 can be related to syenite intrusions (see Fig. 3). Peaks in grout consumption and an enlargement in water leakages can be observed in these areas. The fluctuations in the area from B720 to B760 may arise from a flat-lying syenite intrusion in the crown and it is also here seen an enlargement in the mapped water leakages.

The sequence in which the holes are drilled may affect the impact on the water factor. When a water-bearing fracture that crosses several holes is drilled through for the first time, this may drain the fracture and the impact on the following holes might be smaller. It is hard to estimate how much this affect the results, but since a relation can be seen between the mapped water leakages and the water factor, it is assumed that the water factor shows water where water is present.

The calculation of water and fracturing values (Eqs. 1 and 2) for the holes was created in such a way that it should take into account that a large single sample value should count more than a small value. This is because it is expected that a rock with a large single water or fracturing value would consume more grout than a small value, since it is expected that there is more water in the rock (larger crack) or a more distinct fracture. The increasing impacts, seen in Fig. 16, on the water factor through the tunnel may be a result of lower incoming water pressure on the tunnel rig caused by the increasing length of water pipes and the tunnel gradient which makes B667 32 m higher than B1224.

There are many factors that must be taken into account when evaluating the grout consumption and DPI results. For the study described in this paper, the interpretation of the figures consists of a subjective and visual comparison, where it may have been difficult to set cut-offs for where the data were correlating or not. Still, some quite significant relationships between the data for different areas have been found. In Figs. 14 and 15, the lines for the water factor, fracturing factor and grout volume are simply fitted over each other. Scale differences can be seen even with the normalized water data. This is not unexpected because of changes in overburden, grouting procedures etc. Still, it is possible to see a relationship between the variables if areas with the special geological conditions are taken into account. The area B735–B667 grouted with micro-cement in Fig. 14 seems to have a scale difference between grout volumes and the water factor, and similar grouting procedures (micro-cement and stop criterion 60/40 bar). This may be because of the increased percentages used for the controlled curing, which have limited the grout volumes.

For future projects, a potentially very interesting use of the DPI during grouting would be to distinguish between the two basically different situations: (1) where there is water and (2) where it is not water. In areas with no water from the DPI, a grouting procedure with lower volumes and end pressure should be used to save time and cement. This might be achieved by using different grouting intensity number (GIN) values (Brantberger et al. 2000; Lombardi and Deere 1993) for different rock conditions based on the DPI factors.

Based on the experience from the Løren project and other Norwegian road tunnels, the MWD/DPI technology is believed to have a great potential for predicting water ingress and other rock mass parameters ahead of the tunnel face. The interpretation of data is, however, still uncertain, and more experience from tunnel projects is required in order for the technology to become a fully reliable, routine prediction tool. It is the intention that data collection for further refinement of this technology will be done at several road projects in the near future.

ACKNOWLEDGMENTS

The authors would like to thank the Norwegian Public Roads Administration for giving them the opportunity to write this article and Mr. Kari Bro, Mr. Patric Mårtenson and Mr. Greger Burman at Rockma Systems AB for providing the data and sharing their knowledge on the subject. In addition, thanks go to Ms. Inger Aakre, Mr. Arild Neby, Mr. Henki Ødegaard and Mr. Alf Kveen for the consultations during the writing process.

REFERENCES

1. Axelsson M, Gustafson G, Fransson Å (2009) Stop mechanism for cementitious grouts at different water-to-cement ratios. Tunn Undergr Space Technol 24(4):390–397

2. Barstad M (2010) E 20 Lørentunnelen-Vanntett utstøpning av dyprenne i Lørentunnelen. The Norwegian Public Roads Administration, teknisk vurdering

3. Boge K (2010) Lørentunnelen. Vanntett utstøpning i dyprenne, Aas-Jakobsen AS

4. Boge K (2012) E-mail on the current developments on groundwater levels and settlements in the Løren tunnel project

5. Brantberger M, Stille H, Eriksson M (2000) Controlling grout spreading in tunnel grouting—analyses and developments of the GIN-method. Tunn Undergr Space Technol 15(4):343—352

6. Carlsvärd C, Wallgren EE (2009) Utvärdering av MWD-teknikens möjligheter att identifiera vatternförande zoner vid Norra Länkenprojektet i Stockholm, Master. Kungliga Tekniska högskolan, Stockholm

7. Henriksen JP, Føyn T (2004) RV 150 Ring 3 Ulven-Sinsen—Forslag til innlekkasjekreav for tunnelene—konsekvenser for omgivelsene. Norconsult AS

8. Humstad T, Høien AH, Hoel JE, Kveen A (2012) Complete software overview of rock mass and support in Norwegian road tunnels. In: Eurock 2012, Stockholm, Sweden, May 2012

9. IBM (2010) IBM SPSS Statistics. IBM SPSS, version 19.0.0.1. IBM, Somers, NY

10. Iversen E (2011) Lørentunnelen—Longitudinal general geology section map

11. Lombardi G, Deere D (1993) Grouting design and control using the GIN-principle. Int Water Power Dam Constr 45(6):15–22

12. Rockma (2011) GPM+ Tunnel. GPM+. Rockma System AB, Skellefteå

13. Schunnesson H (1997) Drill process monitoring in percussive drilling for location of structural features, lithological boundaries and rock properties, and for drill productivity evaluation. Doctoral thesis, Luleå University of Technology, Luleå

14. Segui JB, Higgins M (2002) Blast design using measurement while drilling parameters. Fragblast 6(3–4):287–299

15. Smith B (2002) Improvements in blast fragmentation using measurement

while drilling parameters. Fragblast 6(3–4):301–310

16. Valli J (2010) Investigation ahead of the tunnel face by use of a measurement-while-drilling system at Olkiluoto. Finland Working Report, vol 2010. Posiva Oy, Eurajoki

17. Vianova (2011) Novapoint Tunnel. Novapoint, 18.10 FP2c edn. Vianova Systems AS, Sandvika

Chapter 4

INFLUENCE OF SAMPLE HEIGHT-TO-WIDTH RATIOS ON FAILURE MODE FOR RECTANGULAR PRISM SAMPLES OF HARD ROCK LOADED IN UNIAXIAL COMPRESSION

Diyuan Li[1], Charlie C. Li[2] and Xibing Li[1]

[1]School of Resources and Safety Engineering, Central South University, Changsha 410083, Hunan, China

[2]Department of Geology and Mineral Resources Engineering, The Norwegian University of Science and Technology (NTNU), 7491 Trondheim, Norway

ABSTRACT

Surface-parallel slabbing is a failure mode often observed in highly stressed hard rocks in underground excavations. This paper presents the results of experimental studies on slabbing failure of hard rock with different sample height-to-width ratios. The main purpose of this study was to find out the condition to create slabbing failure under uniaxial compression and to determine the slabbing strength of hard rock in the laboratory. Uniaxial compression tests were carried out using five groups of granite specimens. The mechanical parameters of the sample rock, Iddefjord granite from Norway, were measured on the cylindrical and Brazilian disc specimens. The transition of the failure mode was studied using rectangular prism specimens. The initiation and the propagation of slabbing fractures in specimens were identified by examining the relationship among the applied stress, strain and the acoustic emission. The stress thresholds identified were compared to those reported by other authors for crack initiation and brittle failure. It is observed that the macro failure mode will be transformed from shear to slabbing when the height/width ratio is reduced to 0.5 in the prism specimens under uniaxial compression. Micro σ_1-parallel fractures initiate when the lateral strain departs from its linearity. Slabbing fractures are approximately parallel to the loading direction. Labotatory tests show that the slabbing strength (σ_{sl}) of hard rock is about 60% of its uniaxial compression strength. It means that if the maximum tangential stress surrounding an underground excavation reaches about the slabbing threshold, slabbing fractures may take place on the boundary of the excavation. Therefore, the best way to stop or eliminate slabbing failure is to control the excavation boundary to avoid the big stress concentration, so that the maximum tangential stress could be under the slabbing threshold.

INTRODUCTION

With increasing depths of mining and tunneling projects, it has been observed that at high stresses hard rocks fail more often in slabbing (or spalling) rather than in shear. Spalling refers to a failure process involving extensional splitting cracks (Fairhurst and Cook 1966). According to Ortlepp's description (Ortlepp1997), spalling or slabbing is generally defined as the formation of stress-induced slabs on the boundary of an underground excavation. It initiates in the region of maximum tangential stresses and results in a V-shaped notch that is local to the boundary of the opening. This type of failure is typical in strain burst of hard rocks (Ortlepp 2001). Besides, on the boundary of underground openings, slabbing is also observed in hard rock pillars (Exadaktylos and Tsoutrelis 1995; Martin and Maybee 2000). For example, Martin and Maybee (2000) observed that the dominant failure mode was progressive slabbing and spalling in pillars of some Canadian hard rock mines. It shows that the strength of hard rock pillar is directly related to the pillar width-to-height ratio and pillar failure is seldom observed in pillars where the width-to-height ratio is greater than 2. However, the failure of pillars is different from the failure of laboratory samples. At first, pillar failure is related to the strength of rock masses, but sample failure is related to the strength of intact rock. Second, the strength of hard rock pillars is lower than the uniaxial compression strength [accounting for about $(60 \pm 10)\%$ of σ_c], irrespective of the width-to-height ratio of the pillars. Third, size and end effects influence the failure mode of laboratory samples and mining pillars in different ways to some degree. Martin and Maybee also indicated the influence of the confining stress in short pillars. Because at pillar W/H > 2 the confinement at the core of the pillar increases significantly, the use of Hoek–Brown brittle parameters will be less appropriate. Hence, the empirical pillar strength formulas should be limited to pillar W/H < 2 (namely H/W > 0.5). Recently, the spalling failure of hard rock specimens was numerically modeled by Cai (2008) using FEM/DEM combined numerical tool (ELFEN). It shows that the generation of tunnel surface-parallel fractures and microcracks is attributed to material heterogeneity and the existence of relatively high intermediated principal stress (σ_2), as well as zero to low minimum principal stress (σ_3) confinement. At present, the International Society for Rock Mechanics (ISRM) has appointed a commission on rock spalling (http://www.isrm.net/gca/index.php?id=900) among its eight commissions. The commission is supposed to put forward some suggestions on laboratory procedures required to assess crack initiation (CI) in laboratory tests using strain gauges, LVDTs and AE, and to suggest what to do if only UCS is available and so on. It can be seen that the problem of spalling and slabbing failure has become a new challenge in rock mechanics.

Figure 1 shows a photograph of slabbing failure in a 3-year-old mine drift excavated in quartzite at 1,000 m depth. The roof was exposed when another parallel drift was excavated. The slabbing failure, in many cases, is shown as densely spaced "onion-skin" fractures or slabs in highly stressed rocks after excavation. The spacing of the stress-induced fractures depends on the magnitude of rock stresses and the strength of rock, as well as the material heterogeneity (Cai 2008). On the one hand, the present studies on slabbing failure of hard rocks mainly concentrate on the description of slabbing phenomenon and how to control slabbing failure in situ (Dowding and Andersson 1986; Fang and Harrison 2002; Martin et al. 1997). On the other hand, most studies are concerned with shear failure of hard rock (Bieniawski 1967a, b, c; Brady and Brown 2004; Lockner et al. 1991; Mogi 2007; Moore and Lockner 1995; Savage et al. 1996) except for a few studies on splitting failure (Holzhausen and Johnson 1979; Horii and Nemat-Nasser 1986; Li Chunlin 1995; Wong et al.2006) in the laboratory compression tests.

Figure. 1: Slabbing failure in the roof of a 3-year-old mine drift excavated in quartzite at 1,000 m depth. The roof was exposed when another parallel drift was excavated.

Both the classic Mohr–Coulomb criterion and the empirical Hoek–Brown criterion are essentially for the shear failure of rock. These criteria are suitable when the confinement pressure is big enough to create shear failure. Under uniaxial compression, both shear failure criteria cannot be applied when the failure is splitting (a kind of extension failure). Stacey (1981) proposed a simple extension strain criterion for fracture of brittle rock. He points out that when the total extension strain in the rock exceeds a critical value, the extension fracture of brittle rock will initiate. However, the critical value of extension strain of brittle rock is difficult to find out. The criterion is not widely used in engineering even though the formula is very simple.

To better understand the failure process before peak strength in hard rock, Eberhardt (1998), Eberhardt et al. (1998, 1999) did a number of experiments to identify and characterize the brittle fracture process by uniaxial compression testing of pink Lac du Bonnet granite in Canada. The shape of the specimens is cylindrical and the height-to-diameter ratio is approximately 2.25. The crack closure (σ_{cc}), crack initiation (σ_{ci}), crack damage (σ_{cd}) and peak strength (σ_{UCS}) in the stress–strain curves were identified by strain gauges and acoustic emission (AE) techniques. However, the failure mode of hard rock under confinement, the end effect and the size effect in the specimens have not been discussed in those studies. The compression failure of concrete columns and the size effect on compression strength of a columns have been carefully studied by Bazant et al. (Bazant et al. 1999; Bazant and Xiang 1996; Bazant and Xiang 1997). It is well known that the compression strength of a column will change from material strength (σ_{UCS}) to buckling strength when the slenderless is increasing to a critical value. The question is what will happen if the slenderless of rock or concrete column is reduced to a certain small value. Will the failure mode change in the short samples?

The brittle failure of hard rock under compression has been studied extensively since the 1960s, for instance by Cook (1965), Bieniawski (1967a, b, c), Ewy and Cook (1990a, b), Li and Nordlund (1993), Martin and Chandler (1994), Hajiabdolmajid et al. (2002), Wong et al. (2006) and Cai (2008). The methodologies of the studies involve laboratory study, numerical modeling, macroscopic and microscopic observations, site investigation and mathematic deduction. In all the studies, the stress–strain constitutive law and the failure mode of the rock are the most concerning issues. With increasing excavation depth, it has been observed that some fractures are developed parallel to the excavation periphery in hard rock masses (Germanovich and Dyskin 2000; Kaiser and McCreath 1994; Martin et al. 1997; Ortlepp 1997; Ortlepp and Stacey 1994). The slabbing fractures parallel to the excavation periphery in deep underground openings have been studied by some engineers and scholars. For instance, Diederichs (2002, 2007), Diederichs et al. (2004) has paid much attention to the tensile spalling failure of hard rock. He compared two curves between the long-term strength of labotatory samples and the in situ strength of hard rock. He concluded that the in situ strength of hard rock was less than the long-term strength of labotatory samples when the confining stress was relatively low compared with UCS, and spalling/slabbing failure might occur at this confining stress condition. Based on the site observations, we tried to design some laboratory tests to find out the mechanism of slabbing failure and to determine the slabbing strength of hard rock. However, so far, research on this subject is very limited. The objectives of this paper are to investigate the influence of sample height-to-width ratio for the transition of failure mode from shear to slabbing and to determine the slabbing strength of hard rock by laboratory tests. Five groups of laboratory tests were carried out, which include cylinder tests, Brazilian tests and three groups of prism specimens' tests. The mechanical parameters of the sample rock, Iddefjord granite from Norway, were measured on the cylindrical specimens and the disc specimens. The transition of the failure mode was studied using the rectangular prism specimens. AE was monitored during testing to detect the crack initiation and propagation in the samples. Both axial and lateral strains were logged in the tests, so that the stress–strain curves could be obtained.

EXPERIMENTAL TESTS

Specimens

Five groups of specimens from granite blocks taken from a quarry in Iddefjord, Norway, were prepared in the laboratory. The average density of the granite is 2,620 kg/m³. The P-wave velocity ranges from 4,000 to 4,700 m/s. Seah (2006) has done some laboratory tests on Iddefjord granite. The physical and mechanical properties of the rock can also be found in his Ph.D. thesis. The five groups of specimens are:

- Group A: three cylindrical specimens, 50 mm in diameter and 125 mm in length;

- Group B: three Brazilian disc specimens, 50 mm in diameter and 25 mm in thickness;

- Group C: three prism specimens, 50 × 25 × 120 mm, with a height/width ratio of 2.4;

- Group D: three prism specimens, 50 × 25 × 50 mm, with a height/width ratio of 1.0;

- Group E: three prism specimens, 50 × 25 × 25 mm, with a height/width ratio of 0.5.

Groups A and B were used to determine the uniaxial compressive strength (UCS) and indirect tensile strength of the granite. Groups C, D and E were used to investigate the influence of the height/width ratio to the transition of the failure mode. The geometric shapes of the five groups of specimens are shown in Fig. 2. The loading ends of all the cylindrical and prismatic specimens were ground parallel and smooth to minimize the end effects. Axial and lateral strain gauges were mounted on the cylindrical surface for the cylindrical specimens and on both the front and back sides of the prismatic specimens to measure the axial and lateral strain during testing.

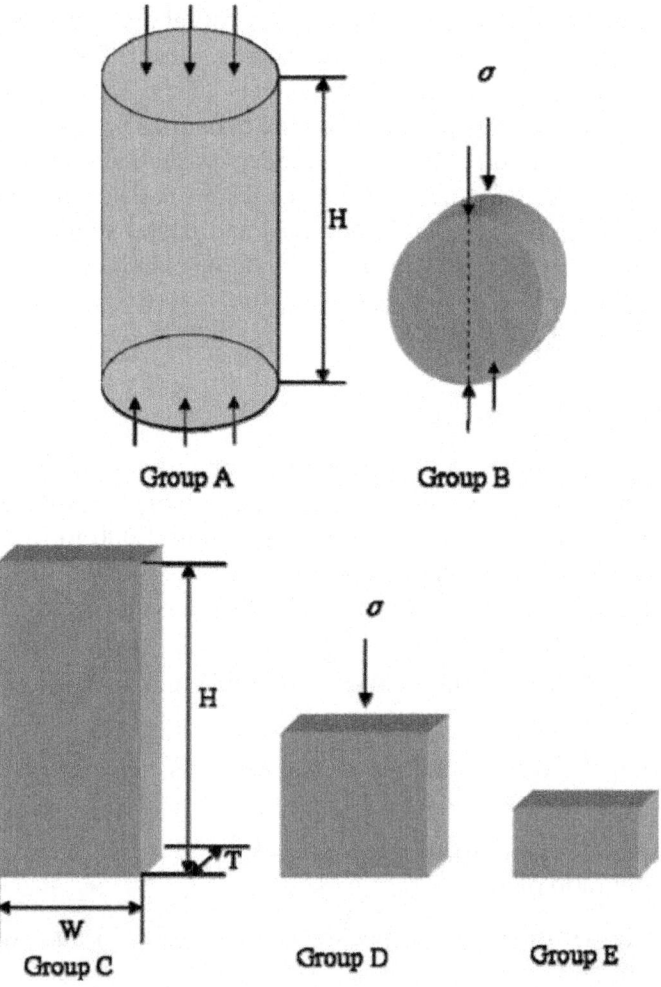

Figure. 2: The geometric shapes of the five groups of Iddefjord granite specimens (H, W, T represent height, width and thickness of the prism specimen, respectively).

Equipments and Testing Procedures

The experiments were carried out on two hydraulic servo-controlled machines, Instron 1346 and Instron 1342, in the Mechanical Testing Centre of Central South University, China. The testing system is controlled by computer and the load and deformation data can be acquired automatically. The uniaxial compression tests of Groups A, C, D and E were carried out on Instron 1346, which has a load capacity of 2,000 KN. The Brazilian tests of Group B were undertaken on Instron 1342, which has a load capacity of 250 KN. The machines are shown in Fig. 3. The specimen's ends were directly in contact with the machine except that a little lubricant was present on the interface under testing.

Figure. 3: The stiff test machines Instron 1342 and 1346 at Central South University.

An AE detecting sensor, a PCI-2 AE data-collecting system and a DH-3817 dynamic-static strain acquisition system were also used for the tests. The threshold of AE trigger level was set to 42 dB for each specimen except A1. The AE signals measured in the sensors were amplified by the gain of 40 dB with pre-amplifiers. The data acquisition rate was set to 0.5 MHz, and a waveform could be measured for every 2 µs. It should be noted that the threshold of AE trigger level in specimen A1 is 39 dB, and it can be seen that the AE counts rate curve oscillates more seriously than A2 and A3 at the beginning of the compression stage because of the influence of background noise. Therefore, the threshold value was set to 42 dB for the other specimens after testing A1. There are two cross-strain gauges on each specimen to measure the axial and lateral strain.

Uniaxial compression tests of group A and Brazilian tests of group B were first conducted to obtain the basic mechanical properties of the Iddefjord granite. The prism specimens in groups C, D and E were uniaxially loaded under compression to study the transition of the failure mode. To study and observe the failure process of the specimens, AE events were monitored and photos of the testing specimens were taken in the process of loading. The loading rate was controlled at 60 KN/min in the beginning stage. When the load reached about 150 KN, the loading was changed from load control to displacement control at a rate of 0.1 mm/min.

Damage Thresholds

It is believed that some stress thresholds exist during the brittle fracturing on the progressive degradation of intact rock under compression. To identify these crack initiation and propagation thresholds in brittle rock, some scholars (Diederichs et al. 2004; Eberhardt et al. 1998) have made contributions to this research field based on axial and lateral deformation measurements, and AE records during laboratory tests. According to the stress–strain characteristics displayed by the axial and lateral strain measurements and AE count rate curves, three damage thresholds were identified during the tests in this paper. By analyzing the laboratory testing data, the axial stress–strain curve, the lateral stress–strain curve, and the logarithmic AE count rate–strain curve can be obtained in a typical figure. Therefore, we define the three damage threshold values in the stress–strain curves for the long specimens in Group A and Group C as follows:

- Stable crack initiation stress (σ_{st}, Point P): the threshold at which the corresponding logarithmic AE count rate curve begins to monotonically increase after oscillating in the beginning stage. It means that the crack growth can be stopped by controlling the applied load.

- Unstable crack development stress (σ_{ust}, Point Q): the threshold at which the corresponding logarithmic AE count rate curve begins to abruptly increase prior to failure. It means that the crack growth would continue even if the applied load were kept constant.

- Slabbing crack initiation stress (σ_{sl}, Point M): the threshold at which the lateral stress–strain response is observed to become nonlinear. It means that the extension slabbing fractures initiate in the brittle rock, and nonlinear increasing of lateral extension strain may contribute to slabbing failure.

Testing Results

Cylinder Compression Tests and Brazilian Disc Tests

Properties of the rock were obtained from the tests of Groups A and B. The properties include the uniaxial compressive strength (σ_c), tensile strength (σ_t), internal friction angle (φ), cohesion (c), Young's modulus (E), Poisson's ratio (v), density (γ) and P-wave velocity (V_p), which are listed in Table 1. The values of φ,and c are obtained by back-calculations from the fracture angle θ and σ_c by using relationships $\varphi=2\theta-90°$ and $c=\sigma c(1-\sin\varphi)/2\cos\varphi$.

Table 1: Basic mechanical properties (mean values) of the Iddefjord granite

σ_c(MPa)	σ_t(MPa)	σ_c/σ_t	θ	φ	c(MPa)	E(GPa)	v	γ(kg/m³)	V_p(m/s)
203.3	8.3	24.5	68°	48°	39.0	51.7	0.19	2620	4,550

The stress–strain curves of the cylindrical specimens in Group A are shown in Fig. 4, and also the curves of the AE count rate (AE counts per second) to the axial strain are shown in a logarithmic scale in the figures. Specimen A1's stress–strain curves in Fig. 4a show, for example, that the Iddefjord granite is very brittle and has a uniaxial compressive strength of 197 MPa. The maximum axial strain is about 3,600 $\mu\varepsilon$ and the maximum lateral extension strain is about 1,800 $\mu\varepsilon$. In the axial stress–strain curve, an initial non-linear behavior owing to crack closure is followed by a linear deformation until the axial stress is about 180 MPa. The lateral strain departs from its linearity when the axial stress is beyond about 130 MPa (point M). After oscillating somehow in the beginning stage of loading, the AE count rate monotonously increases from stress level of 70–180 MPa (zone PQ in the stress–strain curve) and then abruptly increases prior to failure. Similar phenomena can be seen in specimens A2 and A3, as shown in Fig. 4b, c. Figure 5 shows the macro failure fractures of specimens in Group A after testing. All the specimens in Group A failed in shear from the macro point of view.

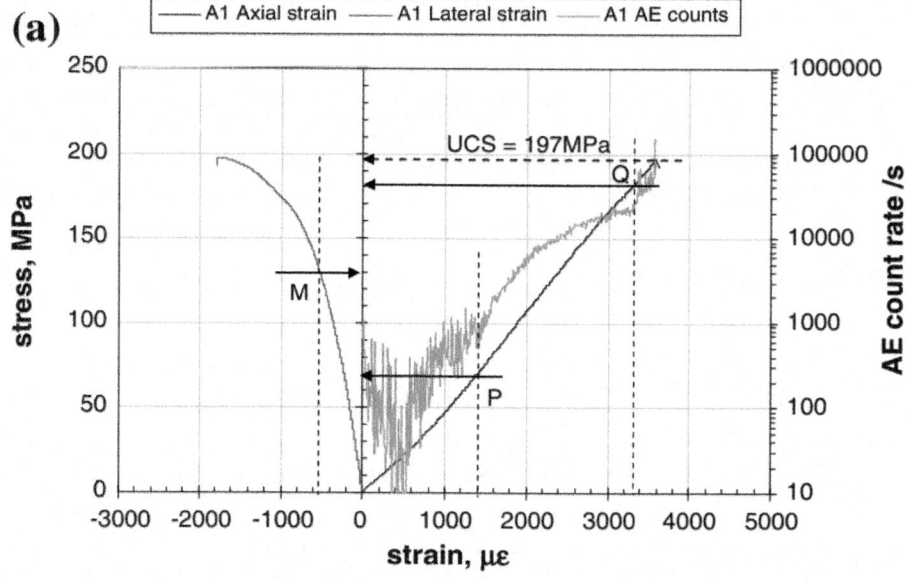

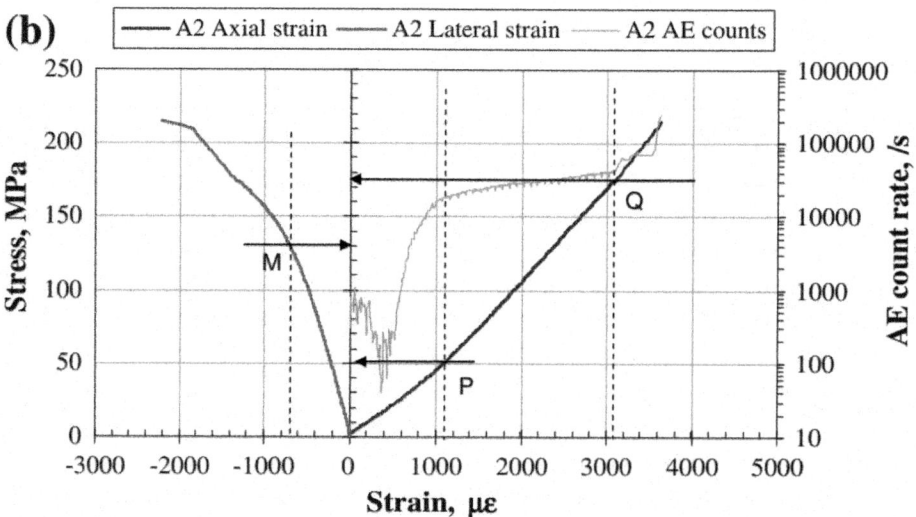

(c)

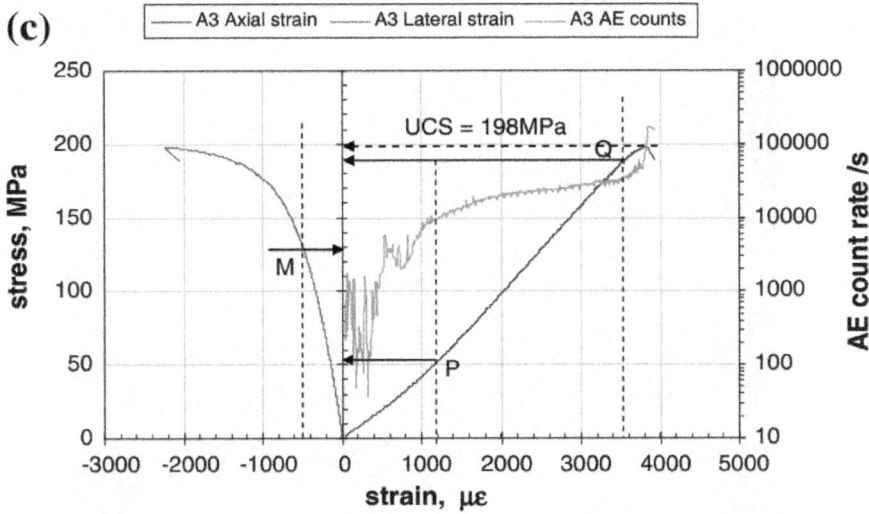

Figure. 4: The stress–strain curves and the AE count rate curves for the cylindrical specimens in group A. **a** Specimen: A1, **b** specimen: A2, **c** specimen: A3.

Figure. 5: The macro failure fractures of specimens in Group A after testing. **a** Specimen: A1, **b** specimen: A2, **c** specimen: A3.

Brazilian disc tests of group B were conducted on Instron 1342. The extension strain and the AE counts were monitored during the tests. The extension strain and the AE counts versus time are shown in Fig. 6. The tensile strength of specimens B1, B2 and B3 are 9.7, 8.7 and 6.3 MPa, respectively. Their maximum extension strains are 510, 410 and 140 με, respectively. The low value of the extension strain for B3 may be due to the lateral strain gauge attached to the specimen not perpendicular to the loading direction. Sporadic AE events occurred until the load was close to the failure point. For instance, it can be seen that intensive AE occurred prior to failure for the specimen B2 in Fig. 6. A similar phenomenon has been observed in specimens B1 and B3. Figure 7 shows the splitting fractures of the three Brazilian disc specimens after testing. The fracture plane of the specimens is approximately along the loading line.

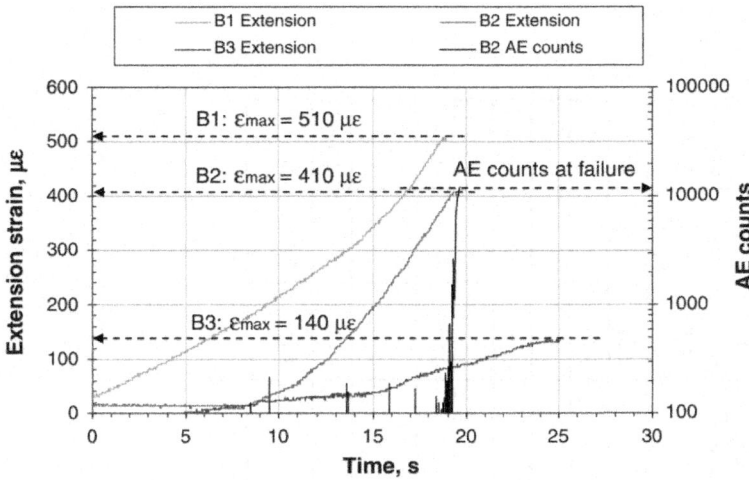

Figure. 6: The extension strain versus time for the Brazilian specimens in Group B and a typical AE counts–time curve for specimen B2.

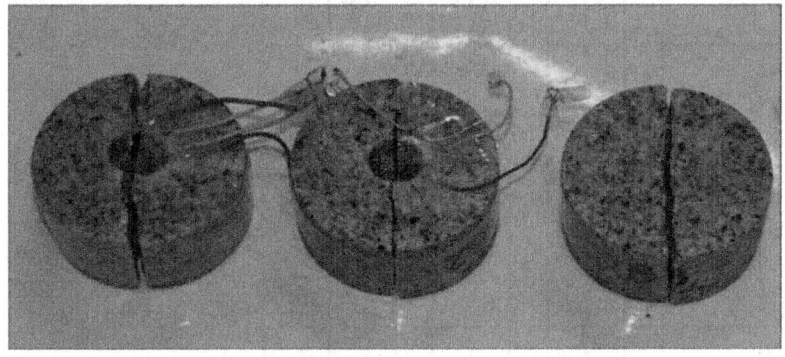

Figure. 7: Splitting fractures of the Brazilian disc specimens in group B after testing.

Compression Tests on the Rectangular Prism Specimens

Uniaxial compression tests on the rectangular prism specimens of groups C, D and E were carried out on Instron 1346. The test results of groups C, D and E are summarized in Table 2.

Table 2: Testing results of the prism specimens in Groups C, D and E

Specimen number	Height,H (mm)	Width,W (mm)	Thickness,T (mm)	Load,P_{max} (KN)	UCS,σ_c (MPa)	Young's modulus,E (Gpa)	Poisson's ratio, v	Fracture angle, θ	Failure mode
C1	120.80	52.20	26.50	236.5	171.0	56.8	0.21	69°	Shear
C2	120.00	51.50	27.60	265.5	186.8	52.7	0.27	70°	Shear
C3	120.70	51.90	26.95	267.0	191.0	50.4	0.19	72°	Shear
D1	51.30	51.90	26.70	320.0	231.0	72.2	0.19	80°	Shear
D2	51.30	51.90	26.70	261.4	188.7	61.6	–	77°	Shear
D3	51.25	51.90	26.70	306.9	221.5	65.4	0.17	78°	Shear
E1	26.40	52.35	26.45	255.1	184.2	57.6	–	85°	Slab-bing
E2	26.40	52.35	26.40	315.6	228.4	62.8	0.12	84°	Hybrid
E3	26.40	52.30	26.45	175.7	127.0	40.1	0.18	88°	Slab-bing

UCS uniaxial compressive strength

The stress–strain curves and also the logarithmic curves of AE count rate for the specimens in groups C, D and E are shown in Figs. 8, 9 and 10, respectively. For the C specimens (H/W = 2.4) in Fig. 8, the average UCS is about 180 MPa, which is about 10% smaller than the UCS of the cylindrical specimens in group A. Both the stress–strain curves and the AE count rate curve are similar to the specimens in group A. The maximum lateral strain of specimen C1 is only about 1,200 με at failure, but for specimens C2 and C3 it is about 2,000 με. The AE count rate of C2 is quite low at low load levels compared to specimens C1 and C3. However, the AE count rate suddenly increases when the load approaches the ultimate failure level. The typical points (point P, Q and M) on the stress–strain curves can be still recognized in the specimens of group C.

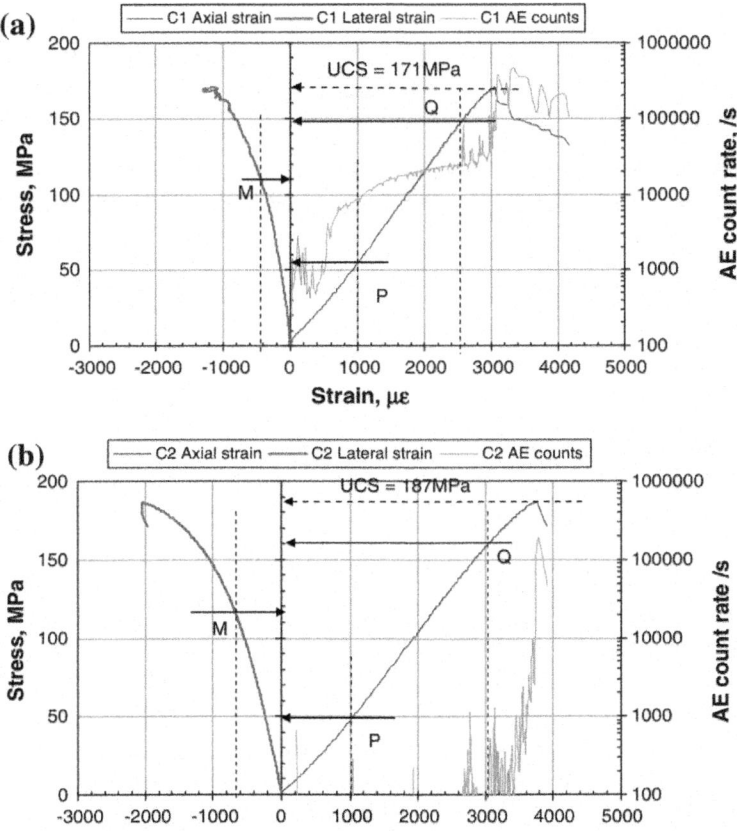

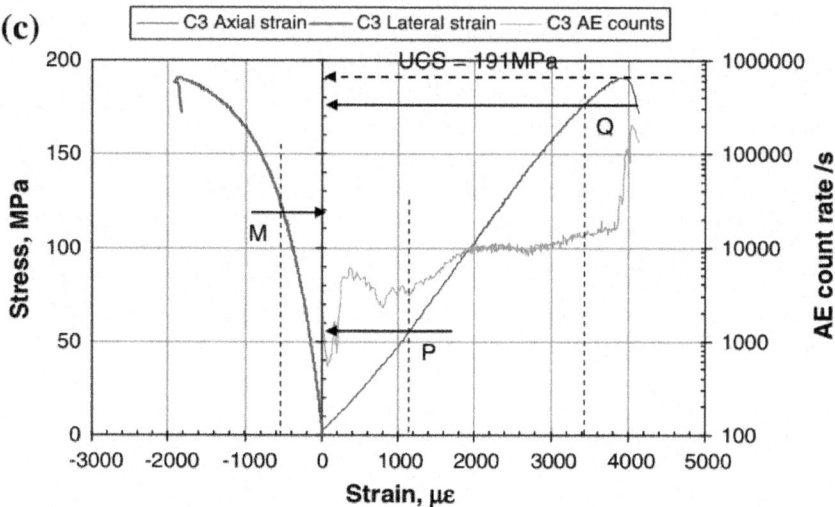

Figure. 8: The stress–strain curves and the AE count rate curves for the prism specimens in group C. **a** Specimen: C1, σ_c = 171 MPa; **b** specimen: C2, σ_c = 187 MPa; **c** specimen: C3, σ_c = 191 MPa.

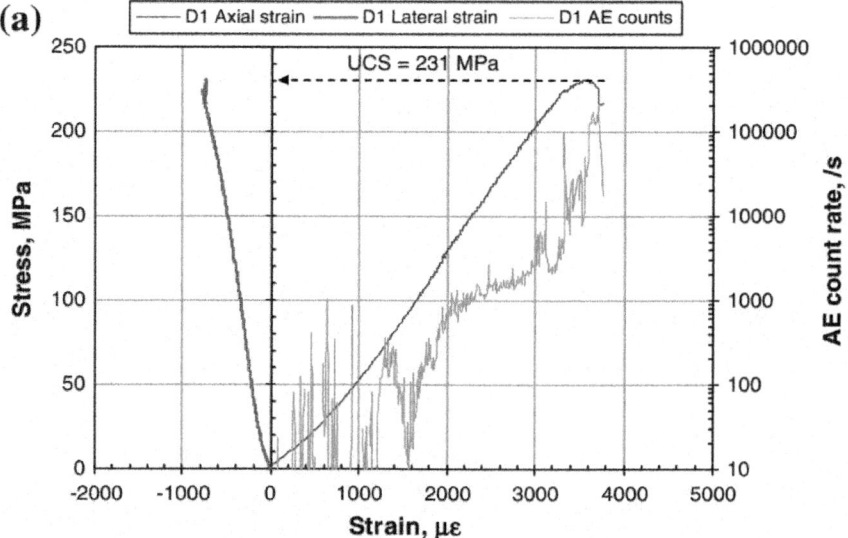

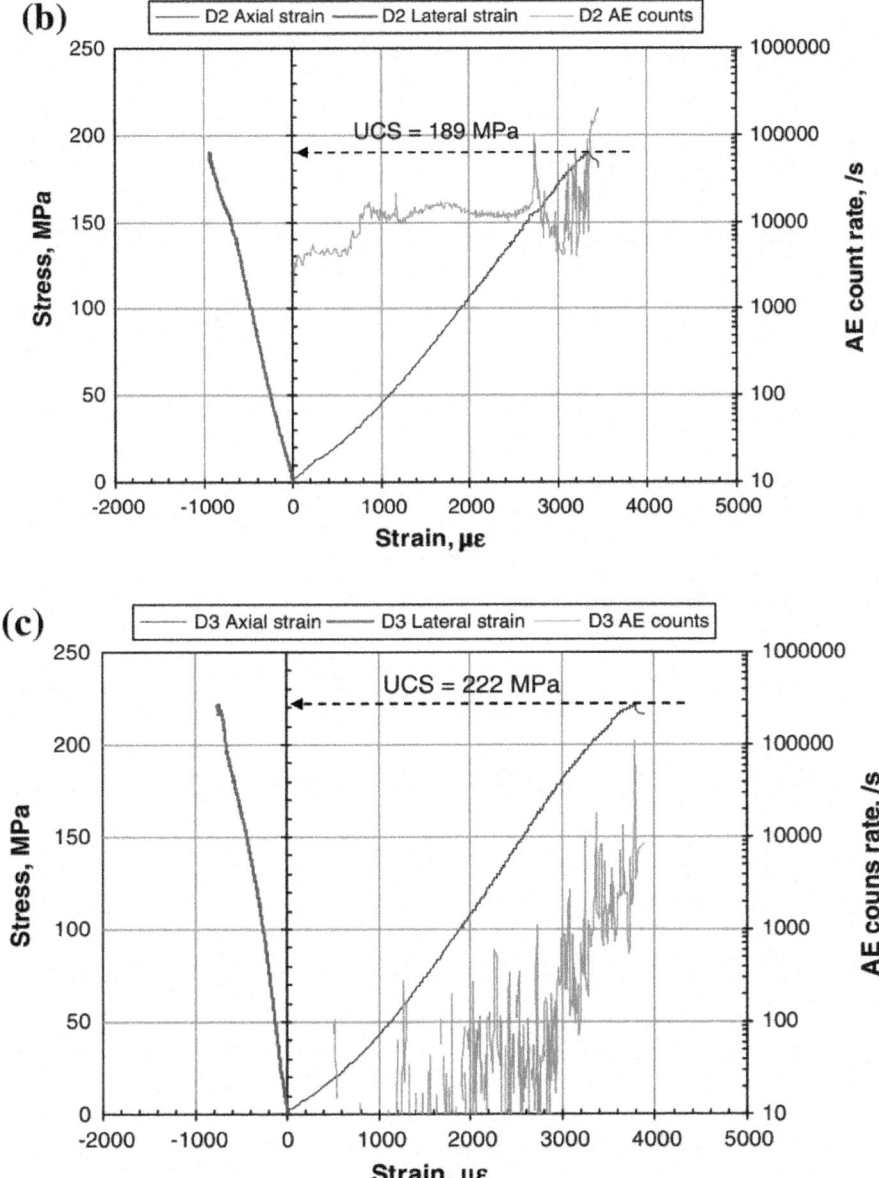

Figure. 9: The stress–strain curves and the AE count rate curves for the prism specimens in group D. **a** Specimen: D1, σ_c = 231 MPa; **b** specimen: D2, σ_c = 189 MPa; **c** specimen: D3, σ_c = 222 MPa.

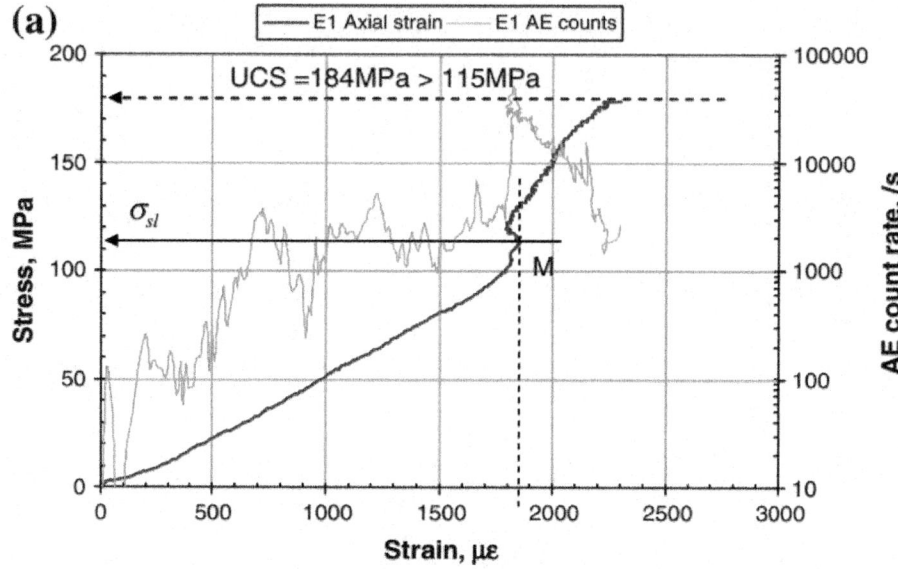

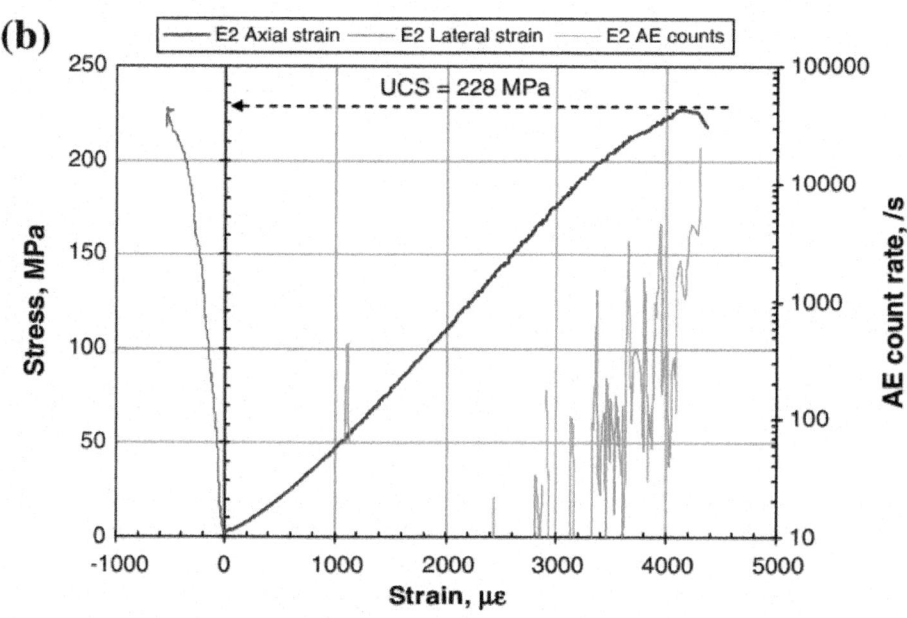

(c)

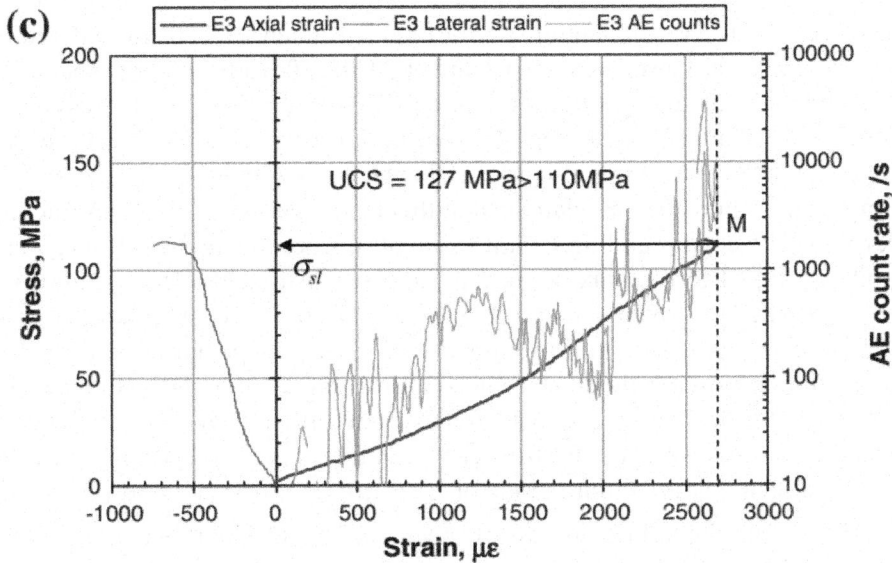

Figure. 10: The stress–strain curves and the AE count rate curves for the prism specimens in group E (σ_{sl} refers to the slabbing strength of the rock). **a** Specimen: E1, σ_c = 184 MPa, σ_{sl} = 115 MPa; **b** specimen: E2, σ_c = 228 MPa; **c** specimen: E3, σ_c = 127 MPa, σ_{sl} = 110 MPa.

Figure 9 shows the testing results of the specimens in group D (H/W = 1.0). Note that the average UCS of the three specimens is about 220 MPa, which is 10% more than the UCS of the cylindrical specimens in group A. Taking specimen D1 as an example in Fig. 9a, a nonlinearity exists in the start portion of the axial stress–strain curve and then it becomes linearly elastic until the stress reaches about 220 MPa. The maximum axial strain is about 3,600 µε, but the maximum lateral strain is only 800 µε. It means that the lateral deformation of D specimens is confined. The lateral stress–strain curve is almost linear even until the final failure. The AE count rate fluctuates when the axial strain is smaller than 1,400 µε and then it monotonically increases until the axial strain reaches about 3,000 µε. After that it becomes nonlinear and an accelerated increase in AE prior to failure is observed. Similar phenomena occur in specimens D2 and D3, as shown in Fig. 9b, c, except that specimen D2 has a relatively larger AE count rate than specimen D1. It may result from the environmental noises during this test.

Figure 10 shows the testing results of the specimens in group E (H/W = 0.5). It seems that these results are much more different than those from the specimens in groups C and D. For example, the UCS of the specimen E1 is about 180 MPa, but it is seen that the axial strain departs from its linearity when the stress reaches about 110 MPa. At this stress level, the AE count rate increases suddenly. The lateral strain of the specimen E1 is disturbed and not listed in Fig. 10a. The curves of specimen E3 (Fig. 10c) are similar to specimens E1. The UCS of the specimen E3 is only 127 MPa. The axial stress–strain curve of specimen E3 also departs from its linearity when the stress reaches about 110 MPa, and at this stress level the AE count rate increases greatly. The strain gauges may become damaged at this load level when slabbing fractures develop. The corresponding lateral strain at this point is about 520 $\mu\varepsilon$, which is almost equal to the maximum extension strain from the Brazilian tests in group B. The UCS of specimen E2 is about 220 MPa. The stress–strain curves of E2 are almost linear prior to failure. The curves of specimen E2 (Fig. 10b) are different from specimens E1 and E3, but similar to the D specimens. The behavior of specimen E2 is considered as an exception from group E, possibly because of the end effects and stress concentration at the corner of the specimen. It can be seen that the failure mode of E2 is different from E1 and E3.

To explain the failure process of specimen E1 and E3 more clearly, the AE counts–time curve and the stress–time curve are shown in Fig. 11. It includes the information of AE counts and applied stress varied with the testing time. Taking the specimen E1 as an example in Fig. 11a, it can be seen that the AE counts increase a lot when the applied stress reaches about 110 MPa. The increase of AE counts during the stress level from 110 to 160 MPa may indicate the initiation and propagation of slabbing fractures in the specimen. Since the fractures are parallel to the maximum loading direction, the specimen can still sustain further stress. It can be found that the slabbing fractures occurred in the specimen E1 from the final failure mode. Similar phenomena can be also observed in the specimen E3. It can be inferred that the slabbing fractures start to propagate at the stress level of 110–120 MPa. From the viewpoint of AE counts curve, the specimen E2 is also an exception in Group E. The reason may be attributed to the confining pressure due to the end effects, which limit the initiation of extension slabbing fractures in the specimen E2. Therefore, we take specimen E2 as an exception in Group E.

Figure 12 shows the failure modes of all the specimens in Groups C, D and E after testing. The failure mode of the specimens in groups C and D is dominant in shear by observing the final macro fractures. However, some splitting fractures and exfoliation failure occurred in the specimens of groups C and D. For the specimens in group E, which are the shortest among the three groups of prismatic specimens, the fracture is approximately vertical, parallel to the direction of loading force. Specimens E1 and E3 failed in a slabbing manner, as seen in Fig. 12g, i. The slabs are thin and are of approximately equal thickness. The slabbing fractures can form either in the long side or in the short side of the sample. It may be dependant on the heterogeneity of the rock samples. Specimen E2 failed at one end, possibly due to stress concentration on the contact surface.

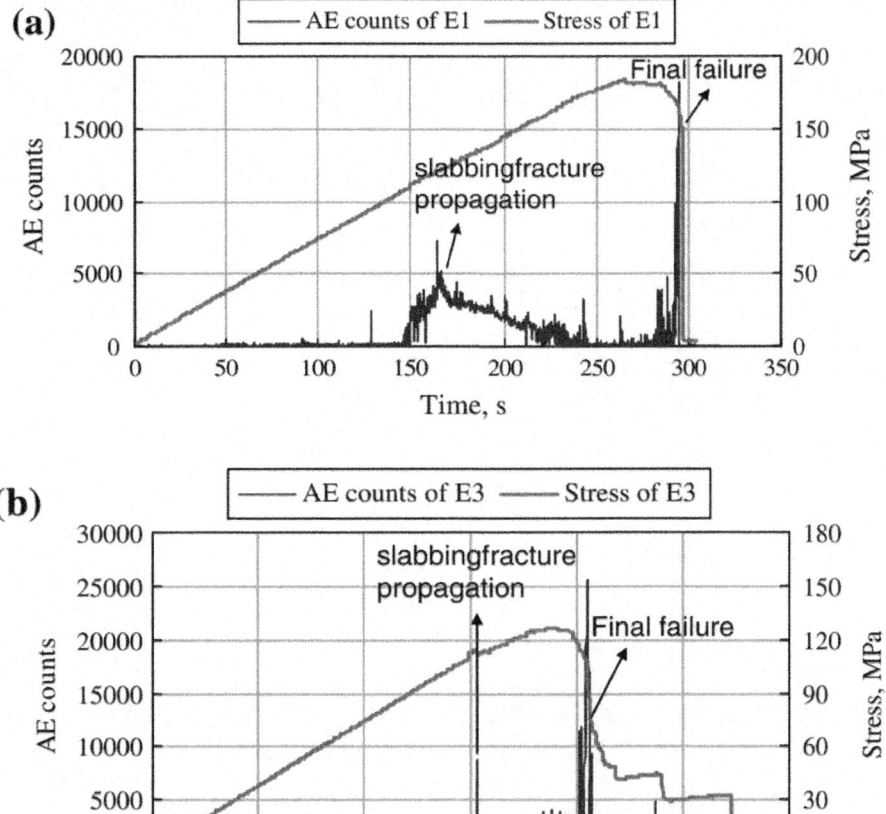

Figure. 11: The AE counts–time curve and the stress–time curve in Group E. **a** Specimen: E1, **b** specimen: E3.

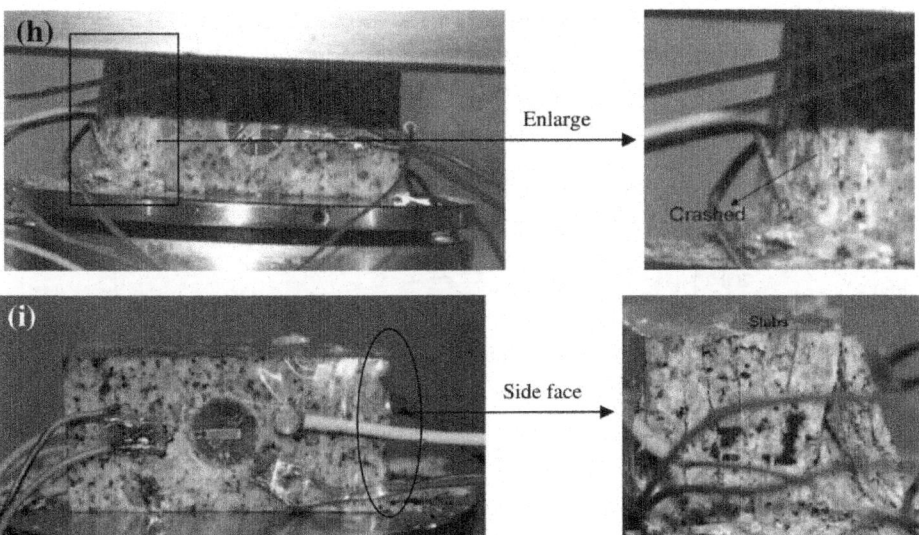

Figure. 12: Failure modes of the prism specimens in groups C, D and E. **a** Specimen: C1, **b** specimen: C2, **c** specimen: C3, **d** specimen: D1, **e** specimen: D2, **f** specimen: D3, **g** specimen: E1, **h** specimen: E2, **I** specimen: E3.

DISCUSSION

The Uniaxial Compressive Strength and the Fracture Angle

The uniaxial compressive strengths (UCS) of all the prismatic specimens (except for specimen E2) in the laboratory tests are plotted versus the H/W ratio in Fig. 13. It is seen that the UCS increases slightly when the H/W ratio decreases from 2.4 (group C) to 1.0 (group D). Shear failure dominates the failure process in C and D specimens. The increase in UCS may be caused by the confinement of the ends in the case of D specimens. The UCS of the E specimens varies over a large range. Two of the three E specimens, E1 and E3, failed obviously in slabbing. Specimen E2 was an exception to Group E, since the failure occurred in a small portion that was ejected from the specimen. It may be the result of the stress concentration caused by mismatching between the specimen ends and the loading platens. The final failure mode of the specimens has been changed from shear at H/W = 1 to slabbing at H/W = 0.5. The drop of UCS for specimens E1 and E3 is possibly due to the change in failure mode. From the curves of specimen E1 and E3 (Figs. 10, 11), it can be seen that the AE count rate increases significantly when the stress reaches about 110 MPa and the axial strain data seems to be invalid or partially invalid afterwards. It can be inferred that when the stress reaches about 110 MPa, the slabbing fractures start to initiate and develop. Since the slabbing fractures are almost parallel to the loading direction, the specimen can still sustain further loading.

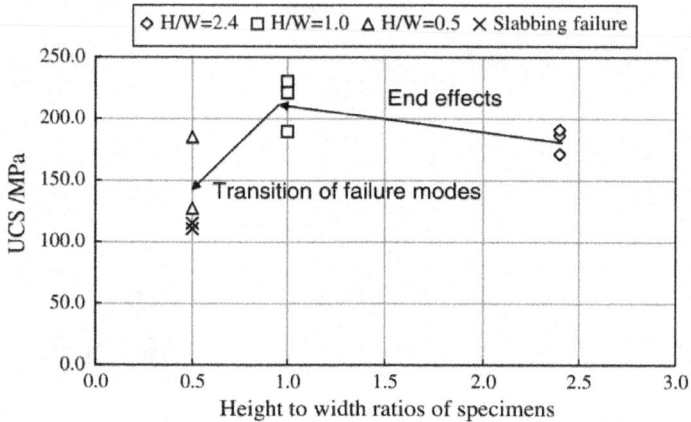

Figure. 13: The variation of the UCS with the height/width ratio for the three groups of prism specimens.

The fracture angle (θ) of a fracture plane is defined as the angle between the normal line to the fracture plane and the loading direction. The fracture angles were measured by the macro fractures of the specimens after testing. The fracture angles of all the compression specimens are plotted in Fig. 14 with respect to the ratio of height to width (H/W).

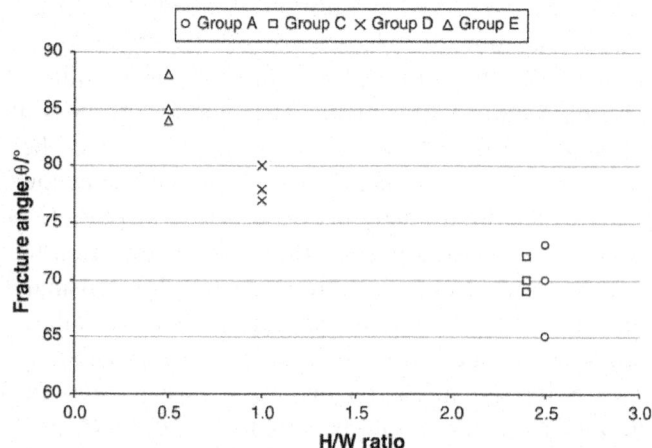

Figure. 14: The variation of the fracture angle with the height/width ratio for the specimens under uniaxail compression.

The fracture angle of the prism specimens changes from about 70° to 86° when the H/W ratio decreases from 2.4 to 0.5. The shear fracture can be clearly observed in the C specimens (H/W = 2.4), but it is not so obvious in the D specimens (H/W = 1.0). The failure mode of the D specimens might be a mixture of shear and slabbing as the result of end effects. The fractures in the shortest specimens (H/W = 0.5) are approximately parallel to the loading direction and almost equal-spaced distributed in the specimens.

Relationship between Stress, Strain and AE Counts

On the stress–strain and AE count rate curves of the long specimens (Groups A and C), three critical points P, Q and M are identified, as shown in Figs. 4 and 8. Point P marks the onset of the monotonically increasing AE count rate, and point Q marks the end of this process. Microcracking initiates and propagates in a stable manner between P and Q. Microcracks begin to coalesce at point Q and propagate in an unstable manner afterward. The stress at P is denoted as σ_{st} (stable crack initiation stress) and the stress at Q is denoted as σ_{ust} (unstable crack development stress). Point M is determined on the stress–lateral strain curve where the lateral strain departs from its linearity. The stress at M is denoted as σ_{sl} (slabbing stress).

Since we define the 'M' threshold as being the lateral strain that departs from its linearity, we can use the secant lateral stiffness as a parameter to determine the 'M' threshold where the lateral stiffness also departs from its linearity. In our analysis, we can obtain the lateral stiffness of specimens by using the stress divided by the lateral strain. These three stress levels can be obtained in Figs. 4 and 8. The value of σ_{st}, σ_{ust} and σ_{sl} for the specimens in Groups A and C are listed in Table 3. The percentages of σ_{st}, σ_{ust} and σ_{sl} to the uniaxial compressive strength are also listed in Table 3. It is seen that they are about 30, 90 and 60% of the UCS, respectively. It is observed that the lateral strain at point M ranges from 420 to 700 $\mu\varepsilon$ and has an average value of 550 $\mu\varepsilon$. It is approximately equal to the maximum extension strain of the rock obtained from the Brazilian disc tests.

Martin and Chandler (1994) and Eberhardt et al. (1998) have put forward some stress thresholds for the crack closure (σ_{cc}), crack initiation (σ_{ci}) and crack damage (σ_{cd}) of hard rock in uniaxial compression tests. The comparisons between our methods and theirs are listed here to show the difference between these thresholds. Martin (1993) suggested one to use calculated crack volumetric strain to identify crack initiation.

Table 3: Data from the critical points on the stress–strain curves of the A and C specimens

Speci-men number	Point P			Point Q			Point M			UCS
	ε_{axial} (με)	σ_{st} (MPa)	σ_{st}/σ_c (%)	ε_{axial} (με)	σ_{ust} (MPa)	σ_{ust}/σ_c (%)	$\varepsilon_{lateral}$ (με)	σ_{sl} (MPa)	σ_{sl}/σ_c (%)	σ_c (MPa)
A1	1,400	70	36	3,280	180	91	550	130	66	197
A2	1,100	50	23	3,100	175	81	700	130	60	215
A3	1,200	52	26	3,550	188	95	460	128	65	198
C1	900	48	28	2,600	148	87	420	105	61	171
C2	1,000	50	27	3,050	160	86	620	115	61	187
C3	1,150	55	29	3,400	176	92	550	120	63	191
Average	1,142	55	29	3,163	171	89	550	121	63	193

For a uniaxially loaded sample, crack volume is determined by subtracting the linear elastic component of the volumetric strain, given by:

$$\varepsilon_{\text{Velastic}} = \frac{1 - 2v}{E} \sigma_{\text{axial}} \tag{1}$$

where E and v are the elastic constants, from the volumetric strain calculated from the measured axial and lateral strain, given by:

$$\varepsilon_{\text{V}} = \varepsilon_{\text{axial}} + 2 \cdot \varepsilon_{\text{lateral}} \tag{2}$$

The remaining volumetric strain is attributed to axial cracking, i.e.,

$$\varepsilon_{\text{V crack}} = \varepsilon_{\text{V}} - \varepsilon_{\text{V elastic}} \tag{3}$$

Therefore, the volumetric strain and the calculated crack volumetric strain can be plotted with applied stress in the stress–strain curves. Martin (1993) defines crack initiation as the stress level at which dilation (i.e., crack volume increase) begins in the crack volume plot. Taking the typical cylindrical specimen A3 and rectangular prism specimen C3 as examples, the stress–strain curves, the AE count rate curve and the stress thresholds are plotted in Figs. 15 and 16, respectively.

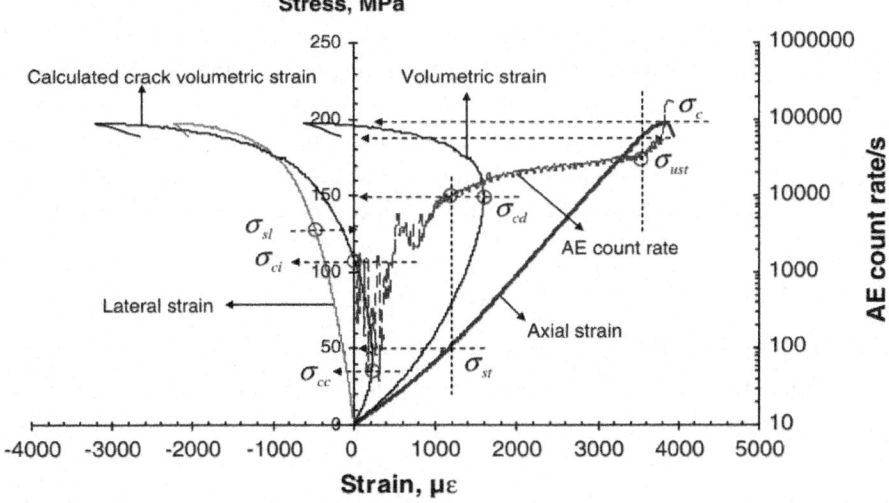

Figure. 15: Determination of stress damage thresholds by Martin's method and ours for the typical cylindrical specimen A3.

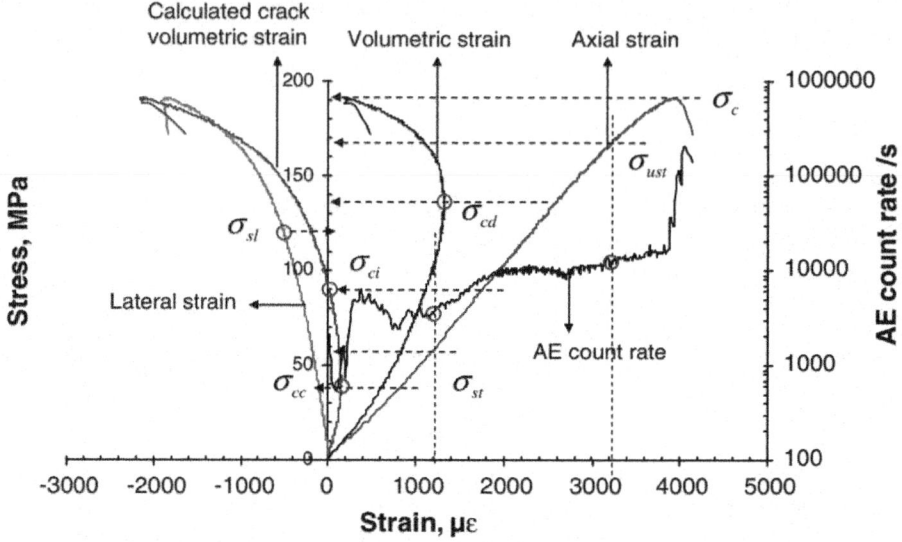

Figure. 16: Determination of stress damage thresholds by Martin's method and ours for the typical rectangular prism specimen C3.

In the Figs. 15 and 16, five curves are shown to determine the stress thresholds at the six points. For example, the thresholds of σ_{cc}, σ_{ci}, and σ_{cd} are determined by the calculated crack volumetric strain curve and the volumetric strain curve by Martin's method. The thresholds of σ_{st}, and σ_{ust} are determined by the AE count rate curve where the AE counts increase monotonically. The threshold of σ_{sl} is determined by the lateral strain curve where it departs from its linearity. It can be seen that these thresholds have some difference in the two figures. The relationship of these thresholds can be described by the following inequality:

$$\sigma_{cc} < \sigma_{st} < \sigma_{ci} < \sigma_{sl} < \sigma_{cd} < \sigma_{ust} < \sigma_{ucs} \qquad (4)$$

where σ_{ucs} is the uniaxial compressive strength.

In the studies of Eberhardt et al. (1998) and Diederichs et al. (2004), the traditional strain measurement method has been used to determine the damage threshold as σ_{cc}, σ_{ci} and σ_{cd}. Meanwhile, these authors have also used the AE measurement result as an important parameter to identify these thresholds. According to the study of Eberhardt et al. (1998), the damage thresholds of σ_{ci} and σ_{cd} were basically determined by the volumetric stiffness and then validated through acoustic emission analysis. If we use this method on the specimen A3 in our tests, the volumetric stiffness versus the applied stress can be obtained and shown in Fig. 17. The damage thresholds can be derived from Fig. 17, where σ_{cc} = 38 MPa, σ_{ci} = 60 MPa and σ_{cd} = 146 MPa. Then the AE event counts, duration and ring down counts versus the axial stress can be drawn to help find out these damage thresholds. Therefore, the acoustic emission measurement is an additional tool for determining these damage thresholds by Eberhardt et al. (1998). It can be seen that the inequality: $\sigma_{cc} < \sigma_{st} < \sigma_{ci} < \sigma_{sl} < \sigma_{cd} < \sigma_{ust} < \sigma_{ucs}$ is still qualified by this method. Note that the damage threshold of σ_{ci} becomes less than the value obtained by Martin's method, and it is almost equal to the value of σ_{st} by our method.

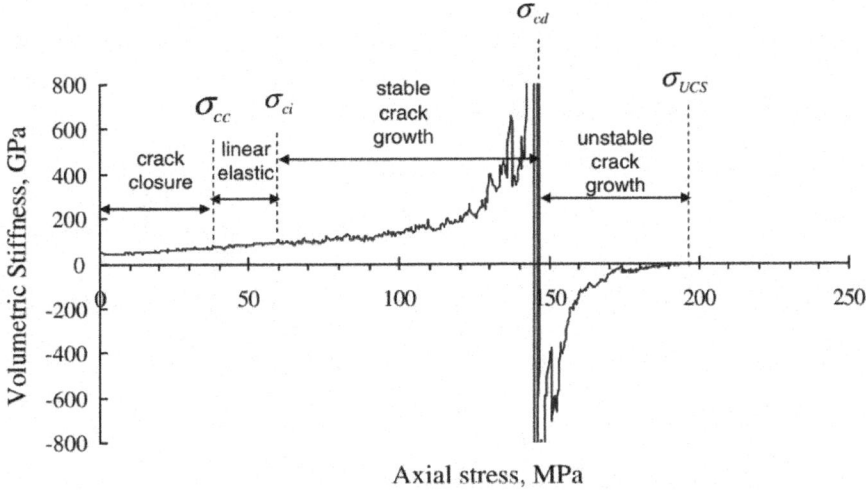

Figure. 17: Determination of the stress damage thresholds by volumetric stiffness with Eberhardt's method in the compressive failure process of specimen A3.

However, some of these damage thresholds are difficult to obtain in the short specimens such as in Group D and E. For example, the critical point M (the threshold of σ_{sl}) did not appear on the stress–strain curves of D specimens (H/W = 1.0). It seems that the lateral extension strain is almost linear prior to failure. The UCS of D specimens is higher than the UCS of C specimens. This may be due to the end effects in the specimens. The strength of the shortest specimens in group E (H/W = 0.5) is not as high as expected, from the view point of the end effects. For example, two of the three specimens have low uniaxial compression strength (specimens E1 and E3). The strength of E1 is about 184 MPa, but at the stress level of 115 MPa the AE count rate has an abrupt jump (Figs. 10a, 11a). It probably means that at this stress level, the slabbing fractures begin to develop in the specimen. The same phenomenon also took place on specimen E3 as shown in Figs. 10c and 11b.

Failure Mode and the Slabbing Strength of Hard Rock

In this study, slabbing failure was observed in the shortest specimens (group E). The failure modes in the prismatic specimens are sketched in Fig. 18. On the one hand, slabbing fractures can be formed in short prismatic specimens under uniaxial compression. Slabs are created parallel to the direction of the maximum compressive load. On the other hand, by carefully observing the shear fractures in the long specimens, it was found that a great number of microfractures parallel to the loading direction existed in the specimens. Since the specimens are long, the microfractures cannot penetrate the samples to form thin slabs easily toward the center of the sample, but finally coalesce together and form a shear band. It means that the fractures will propagate parallel with the loading direction and, finally, form slabbing fractures in short specimens, but will end up as shear fractures in long specimens.

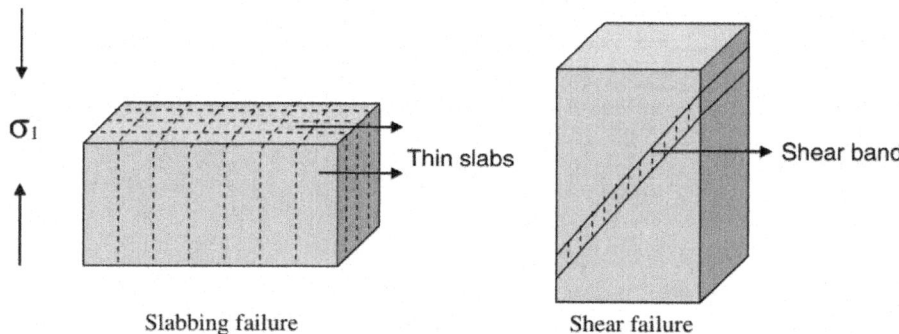

Figure. 18: The failure modes of rectangular prism specimens of hard rock.

When slabbing failure occurs, the corresponding stress is defined as the slabbing strength of the rock. According to the laboratory study, the slabbing-failed specimens E1 and E3 have lower strengths than C and D specimens. The slabbing strength of Iddefjord granite is about 120 MPa. This value is approximately equal to the stress level when the lateral stain departs from its linearity for the long specimens. The slabbing strength of the Iddefjord granite is about 60% of its UCS value for standard specimens.

The observations of the Atomic Energy of Canada Limited (AECL's) Mine-by Experiment support the above claim on the slabbing strength of hard rock (Martin 1997). It was observed in the Mine-by Experiment that the brittle spalling failure initiated when the maximum tangential stress on the boundary of the tunnel reached 120 MPa. The mean uniaxial compressive strength of Lac du Bonnet granite was 212 MPa. The onset of brittle failure (spalling or slabbing) occurred, therefore, at a stress level of 120/212 = 0.56 UCS. This 120 MPa value was also confirmed by Read (2004), by excavating tunnels with various shapes and various orientations relative to the in situ stress state at the 420 Level in the experiment mine. It was reported that the in situ stress magnitudes were $\sigma_1 = 60 \pm 3$, $\sigma_2 = 45 \pm 4$ and $\sigma_1 = 11 \pm 2$ MPa at the 420 Level. The in situ stresses are very high and after tunnel excavation the stress concentrations around the tunnel will lead to approximately uniaxial compression on the boundary of the underground excavation. The loading conditions are similar to the laboratory tests; then the spalling or slabbing fractures begin to propagate and finally form a V-shaped notch at the stress concentration place around the excavation. The stress level for creating slabbing fractures is about $(60 \pm 5)\%$ of the UCS of Lac du Bonnet granite in the site.

The induced stresses and the stress path are both related to the spalling failure of hard rock. On the one hand, when excavating a tunnel in highly stressed hard rocks, the maximum tangential stress surrounding the tunnel can be predicted by the Krisch's equation: $\sigma_{tan} = 3\sigma_1 - \sigma_3$. If the in situ stresses are very high, then the maximum tangential stress may reach the slabbing strength of the rock. On the other hand, when the tunnel is excavated, the radial stress will be near zero. It means that the stress path is changed and the stress conditions created in the surrounding rock become almost uniaxial compression. Therefore, we can conclude that if the maximum tangential stress surrounding an underground excavation reaches about 60% of UCS value of intact rock, slabbing and spalling failure may take place on the boundary of the excavation. The in situ stresses and the induced stresses play an important role in the formation of slabbing fractures in underground engineering. The best way to stop or eliminate slabbing failure is to control the excavation boundary to avoid big stress concentration, so that the maximum tangential stress could be under the slabbing threshold. Another method in underground mining engineering is to apply backfill technology, which can supply certain confining pressure on the surrounding rock to stop the growing of slabbing fractures.

CONCLUSIONS

By changing the sample height-to-width ratio in the uniaxial compression laboratory tests, slabbing failure is achieved in the short rectangular prismatic specimens. It is found out that the failure mode of hard rock may be transformed from shear to slabbing when the height/width ratio of the prism specimen is smaller than, for example, 0.5 under uniaxial compression. The slabbing strength of hard rock is about 60% of the UCS of the rock. Slabbing fractures are approximately parallel to the loading direction, so that the slabs of hard rock can still sustain some further loading. The initiation and propagation of slabbing fractures under uniaxial compression may occur when the lateral extension strain reaches about the maximum extension strain obtained from the Brazilian disc test.

For long specimens, such as groups A and C, the lateral strain will depart from its linearity at this critical point. The stress thresholds of σ_{st}, σ_{ust} and σ_{sl} are compared with the thresholds of σ_{cc}, σ_{cd} and σ_{ci} in the long specimens. It is found that the relationship between them can be described by the inequality as: $\sigma_{cc} < \sigma_{st} < \sigma_{ci} < \sigma_{sl} < \sigma_{cd} < \sigma_{ust} < \sigma_{ucs}$.

The in situ stresses and the induced stresses play an important role in the formation of slabbing fractures in underground engineering. If the maximum tangential stress surrounding an underground excavation reaches about the slabbing threshold (about 60% of UCS value of intact rock), slabbing and spalling failure may take place on the boundary of the excavation. The best way to stop or eliminate slabbing failure is to control the excavation boundary to decrease the stress concentration. Another way in mining engineering is to use backfilling technology, which can supply certain confining pressure on the surrounding rock to stop the growing of slabbing fractures.

ACKNOWLEDGMENTS

The research presented in this paper was jointly supported by the 973 Program of China (grant No. 2010CB732004), the Natural Science Foundation of China (grant No. 50934006 and 10872218) and the Norwegian Research Council (grant no. 10330422). The first author would like to thank the Chinese Scholarship Council for financial support to the joint PhD study at NTNU, Norway. The authors would like to thank Arild Monsøy at NTNU, Norway, Trond Larsen at SINTEF, Norway, Professor Feng Chen and Dr. Chunde Ma at Central South University, China, for their assistance in the preparation of the specimens and the laboratory testing. Also, the authors express their acknowledgements to the anonymous reviewers for their precious comments.

REFERENCES

1. Bazant ZP, Xiang Y (1996) Compression failure in reinforced concrete columns and size effect, Structures Congress—Proceedings, pp 443–451

2. Bazant ZP, Xiang Y (1997) Size effect in compression fracture: splitting crack band propagation. J Eng Mech 123:162–172

3. Bazant ZP, Kim JH, Daniel IM, Becq-Giraudon E, Zi G (1999) Size effect on compression strength of fiber composites failing by kink band propagation. Int J Fract 95:103–141

4. Bieniawski ZT (1967a) Mechanism of brittle fracture of rock. Part I. Theory of the fracture process. Int J Rock Mech Min Sci 4:395–406

5. Bieniawski ZT (1967b) Mechanism of brittle fracture of rock. Part II. Experimental studies. Int J Rock Mech Min Sci 4:407–423

6. Bieniawski ZT (1967c) Mechanism of brittle fracture of rock. Part III. Fracture in tension and under long-term loading. Int J Rock Mech Min Sci 4:425–430

7. Brady BHG, Brown ET (2004) Rock mechanics for underground mining. Kluwer, Dordrecht

8. Cai M (2008) Influence of intermediate principal stress on rock fracturing and strength near excavation boundaries—insight from numerical modeling. Int J Rock Mech Min Sci 45:763–772

9. Chunlin Li (1995) Micromechanics modelling for stress–strain behaviour of brittle rocks. Int J Numer Anal Methods Geomech 19:331–344

10. Cook NGW (1965) The failure of rock. Int J Rock Mech Min Sci 2:389–403

11. Diederichs MS (2002) Stress induced damage accumulation and implications for hard rock engineering. In: Hammah R, Bawden W, Curran J, Telesnicki M (eds) Mining and tunnelling innovation and opportunity. Proceedings of the 5th North American Rock Mechanism Symposium and the 17th Tunnelling Association of the Canada Conference, Toronto 1:3–12. Univ. Toronto Press, Toronto

12. Diederichs MS (2007) The 2003 Canadian Geotechnical Colloquium: Mechanistic interpretation and practical application of damage and spalling prediction criteria for deep tunnelling. Can Geotech J 44:1082–1116

13. Diederichs MS, Kaiser PK, Eberhardt E (2004) Damage initiation and propagation in hard rock during tunnelling and the influence of near-face stress rotation. Int J Rock Mech Min Sci 41:785–812

14. Dowding CH, Andersson CA (1986) Potential for rock bursting and slabbing in deep caverns. Eng Geol 22:265–279

15. Eberhardt E (1998) Brittle rock fracture and progressive damage in uniaxial compression, Ph.D. thesis, Department of Geological Sciences. University of Saskatchewan, Saskatoon, Canada, p 334

16. Eberhardt E, Stead D, Stimpson B, Read RS (1998) Identifying crack initiation and propagation thresholds in brittle rock. Can Geotech J 35:222–233

17. Eberhardt E, Stead D, Stimpson B (1999) Quantifying progressive pre-peak brittle fracture damage in rock during uniaxial compression. Int J Rock Mech Min Sci 36:361–380

18. Ewy RT, Cook NGW (1990a) Deformation and fracture around cylindrical openings in rock. I. Observations and analysis of deformations. Int J Rock Mech Min Sci 27:387–407

19. Ewy RT, Cook NGW (1990b) Deformation and fracture around cylindrical openings in rock-II. Initiation, growth and interaction of fractures. Int J Rock Mech Min Sci Geomech Abstr 27:409–427

20. Exadaktylos GE, Tsoutrelis CE (1995) Pillar failure by axial splitting in brittle rocks. Int J Rock Mech Min Sci Geomech Abstr 32:551–562

21. Fairhurst C, Cook NGW (1966) The phenomenon of rock splitting parallel to the direction of maximum compression in the neighborhood of a surface. In: Proceedings of the First Congress on the International Society of Rock Mechanics, Lisbon, pp 687–692

22. Fang Z, Harrison JP (2002) Development of a local degradation approach to the modelling of brittle fracture in heterogeneous rocks. Int J Rock Mech Min Sci 39:443–457

23. Germanovich LN, Dyskin AV (2000) Fracture mechanisms and instability of openings in compression. Int J Rock Mech Min Sci 37:263–284

24. Hajiabdolmajid V, Kaiser PK, Martin CD (2002) Modelling brittle failure of rock. Int J Rock Mech Min Sci 39:731–741

25. Holzhausen GR, Johnson AM (1979) Analyses of longitudinal splitting of uniaxially compressed rock cylinders. Int J Rock Mech Min Sci Geomech Abstr 16:163–177

26. Horii H, Nemat-Nasser S (1986) Brittle failure in compression: splitting, faulting and brittle–ductile transition. Philos Trans R Soc Lond (Math Phys Sci) 319:337–374

27. Kaiser PK, McCreath DR (1994) Rock mechanics considerations for drilled or bored excavations in hard rock. Tunn Undergr Space Technol Incorporat Trench 9:425–437

28. Li C, Nordlund E (1993) Deformation of brittle rocks under compression-with particular reference to microcracks. Mech Mater 15:223–239

29. Lockner DA, Byerlee JD, Kuksenko V, Ponomarev A, Sidorin A (1991) Quasi-static fault growth and shear fracture energy in granite. Nature 350:39–42

30. Martin CD (1993) Strength of Massive Lac du Bonnet Granite around underground openings, Ph.D. thesis, Department of Civil and Geological Engineering, University of Manitoba, Winnipeg

31. Martin CD (1997) Seventeenth Canadian geotechnical colloquium: the effect of cohesion loss and stress path on brittle rock strength. Can

Geotech J (Revue Canadienne de Geotechnique) 34:698–725

32. Martin CD, Chandler NA (1994) Progressive fracture of Lac du Bonnet granite. Int J Rock Mech Min Sci 31:643–659

33. Martin CD, Maybee WG (2000) The strength of hard-rock pillars. Int J Rock Mech Min Sci 37:1239–1246

34. Martin CD, Read RS, Martino JB (1997) Observations of brittle failure around a circular test tunnel. Int J Rock Mech Min Sci Geomech Abstr 34:1065–1073

35. Mogi K (2007) Experimental rock mechanics. Taylor & Francis, London

36. Moore DE, Lockner DA (1995) Role of microcracking in shear-fracture propagation in granite. J Struct Geol 17:95–114

37. Ortlepp WD (1997) Rock fracture and rockbursts: an illustrative study. South African Institute of Mining and Metallurgy, Johannesburg

38. Ortlepp WD (2001) The behaviour of tunnels at great depth under large static and dynamic pressures. Tunn Undergr Space Technol 16:41–48

39. Ortlepp WD, Stacey TR (1994) Rockburst mechanisms in tunnels and shafts. Tunn Undergr Space Technol 9:59–65

40. Read RS (2004) 20 years of excavation response studies at AECL's underground research laboratory. Int J Rock Mech Min Sci 41:1251–1275

41. Savage JC, Lockner DA, Byerlee JD (1996) Failure in laboratory fault models in triaxial tests. J Geophys Res 101:22215–22224

42. Seah CC (2006) Penetration and perforation of granite targets by hard projectiles. Ph.D. thesis, The Norwegian University of Science and Technology, Trondheim, Norway

43. Stacey TR (1981) A simple extension strain criterion for fracture of brittle rock. Int J Rock Mech Min Sci 18:469–474

44. Wong RHC, Lin P, Tang CA (2006) Experimental and numerical study on splitting failure of brittle solids containing single pore under uniaxial compression. Mech Mater 38:142–159

Chapter 5

SHEETING JOINTS: CHARACTERISATION, SHEAR STRENGTH AND ENGINEERING

S. R. Hencher[1], S. G. Lee[2], T. G. Carter[3], L. R. Richards[4]

[1]Halcrow China Ltd., Hong Kong, China
[2]University of Seoul, Seoul, Korea
[3]Golder Associates, Toronto, Canada
[4]Canterbury, New Zealand

ABSTRACT

Sheeting joints are extensive fractures that typically develop parallel to natural slopes. Embryonic sheeting joints initially constitute channels for water flow and then become the focus for weathering and sediment infill accompanied by progressive deterioration and dilation. Slabs of rock fail along them periodically because of their adverse orientation and long persistence. They are however rough and wavy and these characteristics contribute highly to their shear strength and improve their stability. This paper reviews several landslide case histories and on the basis of these provides guidelines for characterising sheeting joints and determining their shear strength. Engineering options for stabilising sheeting joints in natural and cut slope configurations are then examined with reference to case examples.

INTRODUCTION

Sheeting joints are a striking feature of many landscapes (Figure 1a, b) and they have been studied for more than two centuries (Twidale 1973). They run roughly parallel to the ground surface in flat-lying and steeply inclined terrain and generally occur close to the surface, typically at less than 30 m depth. They can often be traced laterally for hundreds of metres. Most sheeting joints are young geologically and some have been observed to develop explosively and rapidly as tensile fractures in response to unloading (Nichols 1980).

Others are propagated to assist in quarrying using heat or hydraulic pressure (Holzhausen 1989). Their recent origins and long persistence without rock bridges differentiates them from most other joints many of which develop following pre-imposed, geological weakness directions during weathering and unloading as illustrated in Figure 2a, b and discussed by Wise (1964) and Hencher and Knipe (2007). These characteristics of sheeting joints (especially their long persistence and lack of intact rock bridges) also distinguish them from most bedding, cleavage or schistosity-parallel discontinuities.

Figure 1: Sheeting joints in granite: **a** Mt. Bukhansan, near Seoul, Korea; **b** Tuen Mun Highway, Hong Kong

Figure 2: Parallel set traces of proto-joints in the process of developing as full mechanical fractures but maintaining considerable true cohesion **a** Jurassic granite, Bukansan, Seoul; **b** Tuff, Island Road, Hong Kong

DEVELOPMENT OF SHEETING JOINTS

Sheeting joints are common in granite and other massive igneous rocks but also develop more rarely in other rock types including sandstone and conglomerate. Ollier (1975) provides an excellent review of early research and observations on their occurrence and development and Twidale and Vidal Romani (2005) discuss their occurrence specifically in granitic terrain.

Some sheeting joints develop at shallow dip angles, for instance during quarrying, where high horizontal compressive stresses are locked in at shallow depths (Figure 3a). In Southern Ontario, Canada, for example, high horizontal stresses locked in following glacial unloading often give rise to quarry floor heave and pop-up structures accompanied by opening up of pre-existing incipient discontinuities such as bedding planes and schistose cleavage (Roorda et al. 1982). Where there are no pre-existing weakness directions, new sub-horizontal fractures may develop in otherwise unfractured rock. Holzhausen (1989) describes propagation of new sheeting joints under a horizontal stress of about 17 MPa at a depth of only 4 m where the vertical confining stress due to self weight of the rock is only about 100 kPa. The mechanism is similar to a uniaxial compressive strength test where tensile fracture propagates parallel to the maximum principal stress (σ_1). Such exfoliation and tensile development of sheeting joints is analogous to the sometimes explosive spalling and slabbing often seen in deep mines (Diederichs 2003; Hoek 1968).

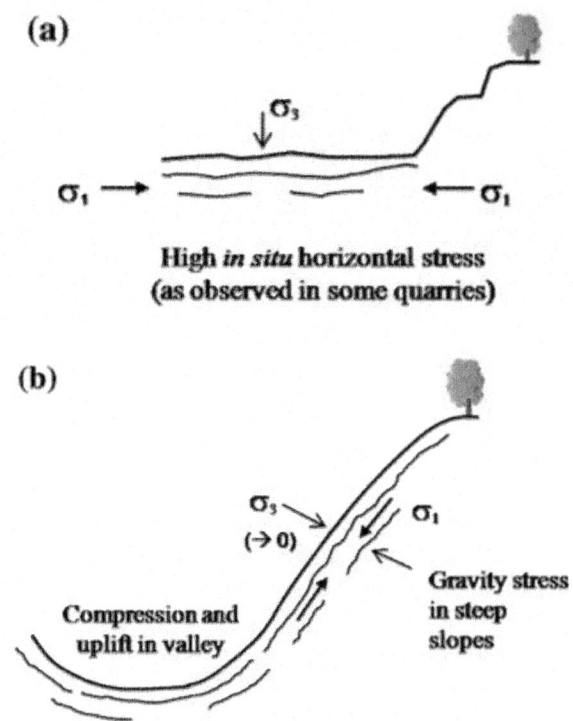

Figure 3: Stress conditions for the formation of sheeting joints **a** in regions with very high horizontal in situ stresses and **b** in steep slopes and relatively strong, unfractured rock.

From a worldwide perspective, however, the joints most commonly recognized as sheeting structures are observed in steep natural slopes. These joints are also thought to develop as tensile fractures where the maximum compressive stress due to gravity is reoriented to run parallel to the slope as illustrated in Figure 3b, demonstrated by numerical models (Yu and

Coates 1970; Selby 1993) and discussed in detail by Bahat et al. (1999). Sheeting joints also develop parallel to the stress trajectories that curve under valleys where there has been rapid glacial unloading or valley down cutting as illustrated in Figure 4. Failure and erosion is a continuing process with the formation of new sheeting joints following the failure of sheet-bounded slabs. Wakasa et al. (2006) calculated an average erosion rate of 56 m in 1 million years from measurements of exposed sheeting joints in granite in Korea which is significantly higher than erosion rates on other slopes without sheeting joints. Whilst many exposed sheeting joints are evidently very recent, others are much older. Jahns (1943) and Martel (2006) note the apparent dissection of landscapes post-dating sheet joint formation. Antiquity is also indicated by preferential and thick weathering as illustrated in Figure 5a, which shows a segment of core through a sheet joint infilled with completely decomposed granite (CDG) with the joint infill material abutting directly against almost fresh rock; Figure 5b shows a similar thick band of highly decomposed granite (HDG) between the walls of a sheeting joint at North Point, Hong Kong. Weathering grades used here are as defined in Geotechnical Engineering Office (1988) and BS 5930 (1999).

Additional evidence for the great age of some sheeting joints is the fact that they can sometimes be observed cutting through otherwise highly fractured rock. Most sheeting joints occur in massive, strong rock and it is argued that if the rock mass had been already highly fractured or weathered then the topographic stresses would be accommodated by movements within the weak mass rather than by initiating a new tensile fracture (Vidal Romani and Twidale 1999). Therefore where sheeting joints are found in highly fractured rock masses, it is likely that they predate the gradual development of the other joints as mechanical fractures during unloading and weathering (Hencher 2006; Hencher and Knipe 2007).

Figure 4: Sheeting joints under valleys **a** Norway; **b** Zambezi River at Batoka Gorge, Zimbabwe. The sheeting joints are in otherwise massive basalt.

Figure 5: **a** Completely decomposed granite in sheeting joint (horizontal borehole for

tunnel). **b** Thick (300 mm) continuous layer of grade IV granite sandwiched between grade II wall rock. Note sharp contacts and rough nature of lower exposed sheet joint surface. Bamboo scaffold pole in foreground.

Some extensive, hillside-parallel joints have many of the characteristics of "true" sheeting joints but owe their geometry instead to the opening up of pre-existing weakness directions such as doming joints in plutonic igneous rock or bedding in sedimentary rock. In this case the pre-existing fracture network defines the hillside shape rather than the other way around (Twidale 1973). The opening up of these pre-existing joint systems is probably largely in response to the same topographic stress conditions that encourage the formation of virginal sheeting joints in massive rock but development may be more gradual. That being so, such joint sets are more likely to retain intact rock bridges between sections of fully developed mechanical fractures than will true sheeting joints and these rock bridges will provide real cohesion, improving overall hillside stability.

GEOMETRY AND OCCURRENCE

Sheeting Joints within the Weathering Profile

Sheeting joint terrain can be regarded as one end of a range of possible weathering profiles where the erosion rate exceeds that of the development of saprolite. They tend to develop in steep sections of hillsides where of course surface erosion is also high. Otherwise there does not appear to be any defining criterion for whether sheeting joints or thicker weathering profiles will dominate across a hillside although it seems clear that once the rock mass becomes significantly fractured or deeply weathered then this will preclude the further development of sheeting joints. They are generally not found in rock underlying thick weathered profiles; the weight of the soil probably constrains their formation (Figure 6). They are however sometimes encountered as relict features within or at the base of thick weathered profiles where they probably predate the weathering. In such situations they may constitute a major hazard for slope stability in that they lack cohesion unlike the surrounding saprolite, they may allow water pressures to develop rapidly and the roughness and waviness that is such an important characteristic of sheeting joints in fresher rock may be ineffective due to weakening of the wall rock.

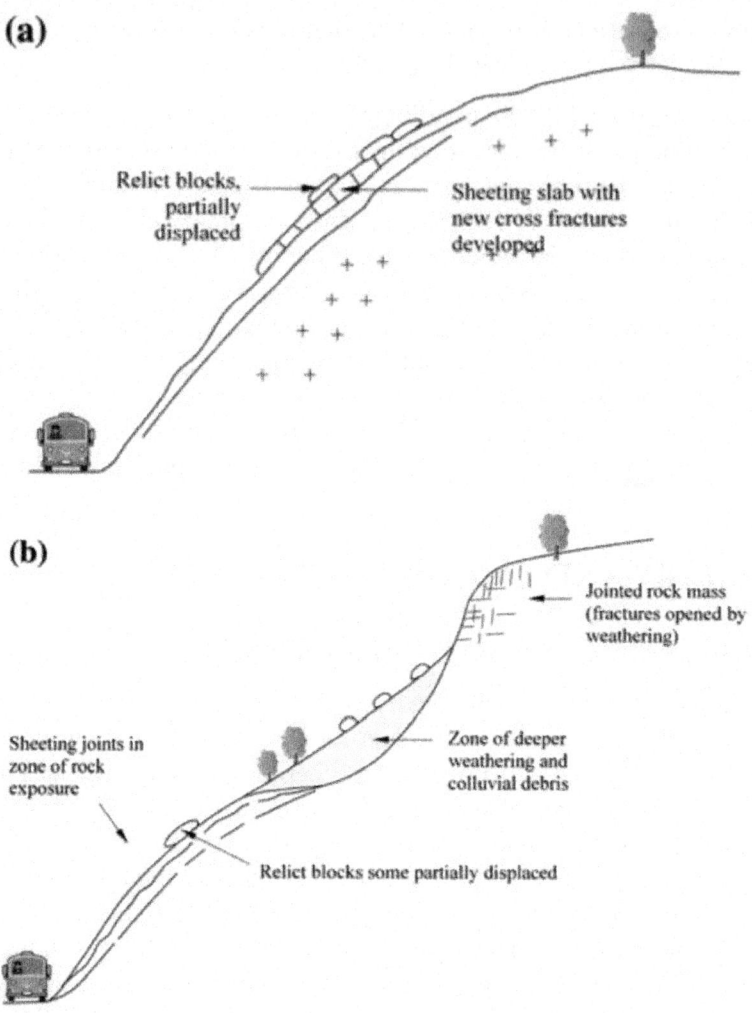

Figure 6: Typical sheeting joint terrains. **a** Sheeting joints dominate landscape; **b** sheeting joints only in areas of rock exposure where erosion rate is higher than accumulation of weathering and colluvial deposits.

General Shape, Occurrence and Relationship to Micro Fractures

Sheeting joints often extend 100 m or more laterally as discrete fractures. Overlying tabular slabs of rock are typically 1–10 m thick; with an observed tendency for slab thickness to decrease the closer they are to the ground surface. The fractures often terminate against pre-existing cross joints or in intact rock (Figure 7). Sometimes, adjacent parts of fractures, perhaps with two or more initiation points, interact forming step-like or "shingle" features as described by Holzhausen (1989). At some locations close microfractures can be observed running throughout the hillside, parallel to the natural slope, instead of discrete fractures defining rock slabs (Figure 8a). Such zones of parallel microfractures probably reflect the overall rotated gravitational stress state in a similar way as do discrete sheeting joints. Elsewhere microfractures develop in a disintegrated zone between and parallel to the sheeting joint walls and have probably developed by exfoliation disintegration away from the original discrete fracture (Figure 8b).

In other locations parallel microfractures in what appears to be sheeting joint terrain may be demonstrated to owe their origins to geological stress conditions predating recent topographic development as for other sets of proto-joints as discussed earlier. During drilling for the Tuen Mun Highway in Hong Kong zones of microfractures were encountered in some areas at the same depths where joints, interpreted as sheeting joints, might have been expected through interpolation from other boreholes (Carter et al. 1998). Detailed petrographic examination however revealed a block-work fabric of microfractures with the main set running parallel to other discrete sheeting joints and to the hillside, whilst others were essentially orthogonal and parallel to sets of cross joints developed elsewhere as mechanical fractures (Figure 9). In some samples all microfractures were seen to be healed with secondary quartz which is taken as an indication of their antiquity.

Clearly microfracturing and sheeting joint development are often related as discussed in some detail by Jahns (1943) and Johnson (1970), but the relationship can be complex. In some situations, pre-existing microfractures can merge and coalesce as "sheeting joints" as Jahns suggests but elsewhere microfractures may develop instead of a discrete fracture and thirdly sheeting joints may develop explosively and essentially independently of any gradual microfracture growth stage. The situation may be particularly difficult to unravel when dealing with igneous plutonic rocks as at the Tuen Mun Highway and as studied by Jahns (1943) and Johnson (1970) and where the "sheeting joints" may exploit a microfracture network originally imposed by cooling stresses following a doming joint pattern (Cloos 1922) and where the cross joints might also owe their origins to the cooling and emplacement process. The evidence of quartz healing in the Tuen Mun Highway samples does suggest that at least some of the microfabric at the Tuen Mun Highway site might well owe its origin to cooling stresses with the quartz healing being the result of late stage throughflow of hot fluids during emplacement (pneumatolysis).

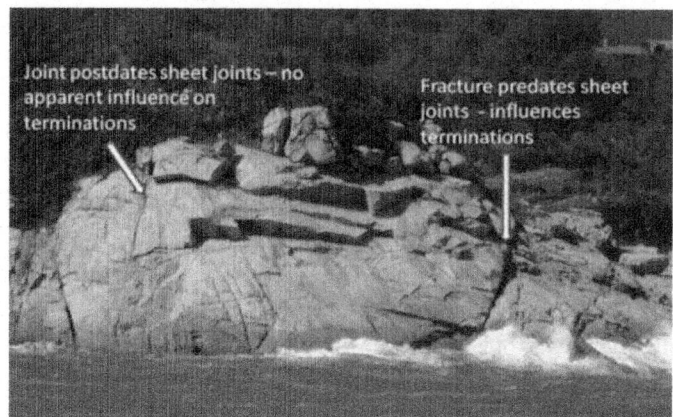

Figure 7: Cross fractures influencing sheet joint development, Shek O, Hong Kong.

Figure 8: Parallel unloading microfractures throughout hillside. **a** Lion Rock Tunnel

portal, Hong Kong. Lens cap for scale. **b** Near Doseonsa temple, nr. Seoul. Note microfracture disintegration is parallel to discrete sheeting joint in background.

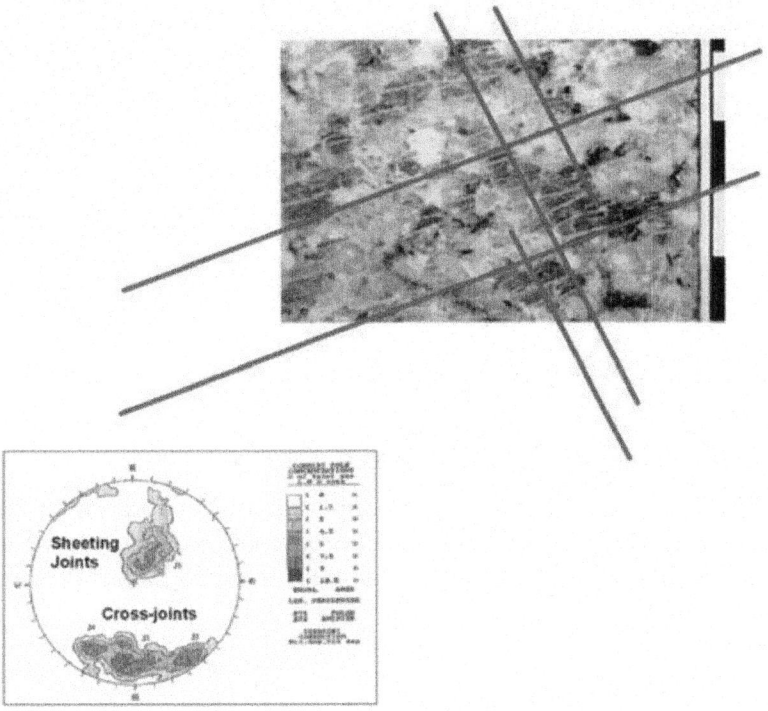

Figure 9: Main set of microfractures dipping to left run parallel to main discrete sheeting joints seen elsewhere in slope (see stereo plot). Cross fractures run roughly parallel to cross joints dipping back into slope and acting as release joints to the sheeting joints.

Surface Characteristics

The flat-lying joints encountered in quarries in areas of high residual horizontal stress and recognized as a type of sheeting or exfoliation joint are described as often smooth by Holzhausen (1989) although some show lineations. Sheeting joints that develop in steeply inclined hillsides where they probably owe their origin to gravitational stresses are typically rough and wavy (Richards and Cowland 1982). They often show broad waves with amplitude of the order of perhaps 1 m over wavelengths of 5–10 m. Smaller roughness features and steps are superimposed on the general waviness as illustrated in Figure 10. Some of the steps are due to intersections with cross joints whilst others may have resulted from different sections of the same sheeting joint interacting and overlapping during propagation as explained by Holzhausen (1989).

Figure 10: Wavy and stepped surface through volcanic tuff, Hong Kong

ENGINEERING CONSIDERATIONS

It is a paradox that sometimes the entire stability of steeply cut slopes in otherwise excellent quality rock can be compromised by the presence of discrete sheeting joints. They are also a major source of landsliding in natural terrain.

Hydrogeology

A newly formed sheeting joint may comprise a near perfectly matching fracture. Nevertheless, the permeability will be higher than the surrounding rock and allow groundwater ingress. In some situations water pressure may reduce effective stresses sufficiently to initiate shear. During shear, rough joints dilate thereby relieving water pressure and halting movement in many situations (Figure 11). Following such transient displacement there will be a period of stasis, during which the walls of the joint may be attacked by weathering, the overlying rock disintegrates and voids that have been opened due to dilation are infilled with sediment transported from upslope. It is common to find depressions on dilated sheeting joints infilled with pockets of sediment, often hard kaolin, and this is interpreted as material that has gradually accumulated over a long period (Hencher 1983; Halcrow Asia Partnership 1998; Parry et al. 2000). Consequently open joints and the presence of sediment infill may indicate that some translational movement has occurred (Hencher 2006). It might take many iterative minor movements from extreme rainfall events before the controlling, wavy first order asperities (Patton and Deere 1970) are overridden and the slab detaches down slope as illustrated by the Leung King Estate, Hong Kong, case example presented below.

Figure 11: Partial movement and dilation of sheeting slabs: typical of early stage of failure **a** Korea, **b** Hong Kong.

The water flow through all joints is tortuous, channelized and localized (Kikuchi and Mito 1993; Hencher2010) and this will be especially true of most sheeting joints with their rough and wavy surfaces. Richards and Cowland (1986) report on a careful instrumentation programme to measure water pressure in a series of sheeting joints. They found that storms resulted in pressure surges through the joints with different joints responding at different times and in different ways during separate storms (Figure 12). Carter et al. (1998) observed that distinctly different behavior occurred between CDG-infilled sheet structures and more broken, less weathered sheet structures, in that the latter showed significant peaky response to rainfall events (with almost complete pressure dissipation occurring in minutes to hours), while the former showed a much delayed decay in pore pressure dissipation after the rainfall event; in some cases taking several days before complete dissipation returned to pre-rainfall pressure head conditions. The implications for design are significant in that peak water pressures do not occur throughout the whole slope at the same time. The tortuosity of drainage paths also has implications for the effective design of landslide preventive drainage measures as discussed later. The drains have to intersect the natural drainage paths if they are to do any good. It must also be anticipated that drainage paths may change with time so that some drains may dry up and others need to be installed.

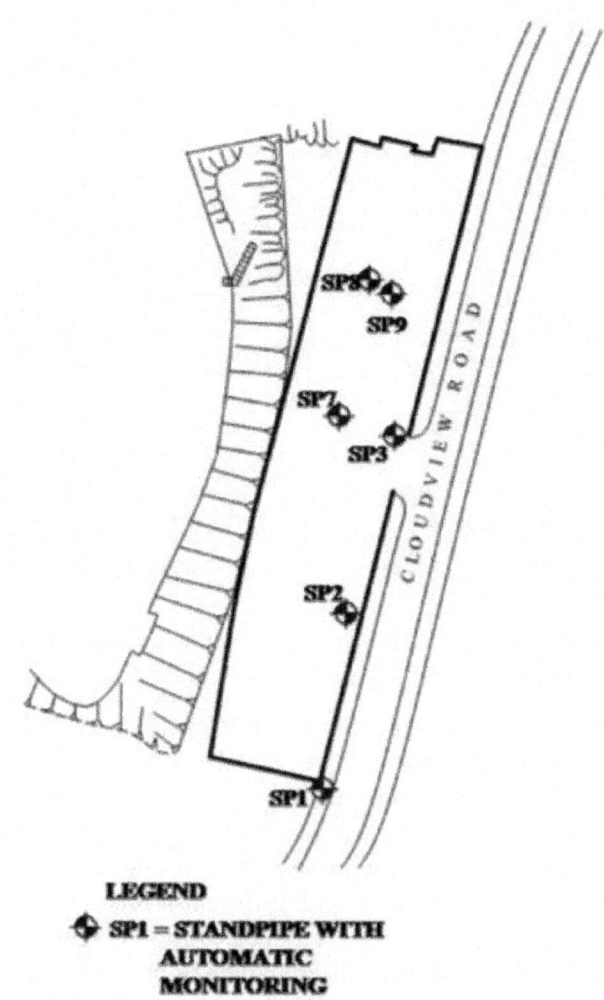

LEGEND

⊕ SP1 = STANDPIPE WITH AUTOMATIC MONITORING

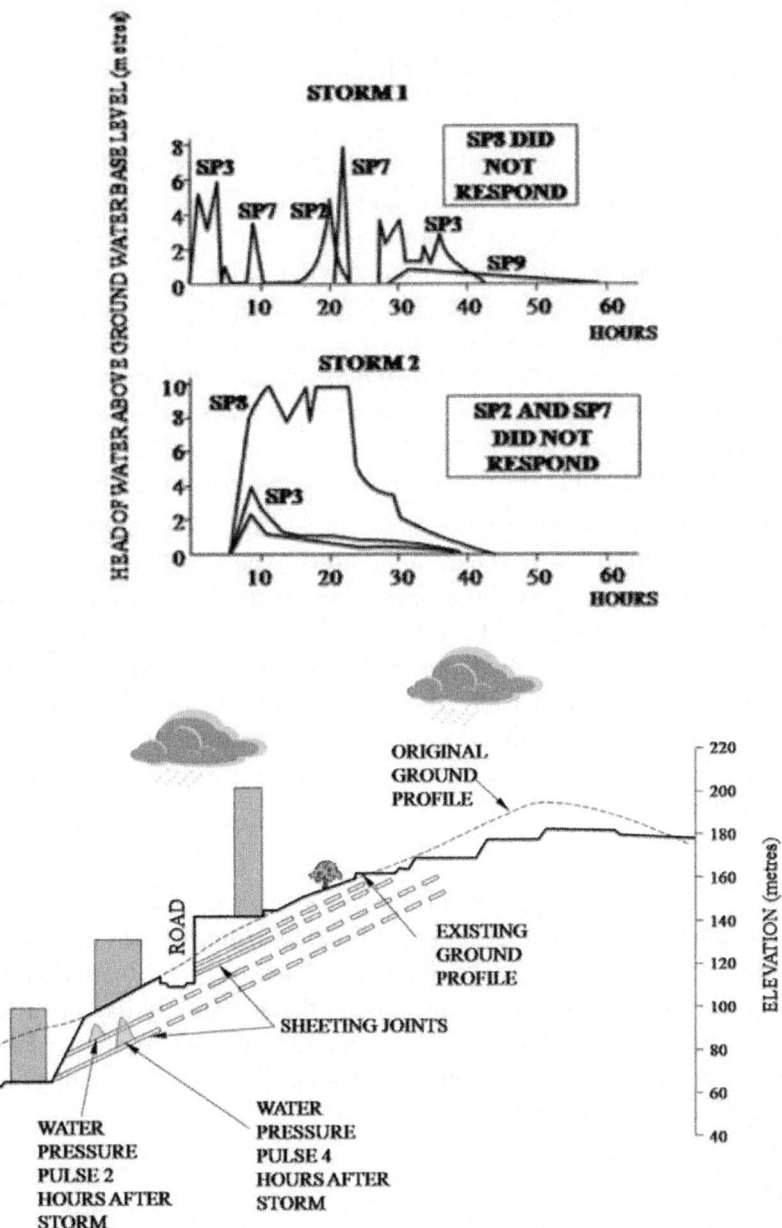

Figure 12: Storm response of piezometers installed in sheeting joints (after Richards and Cowland 1986).

Shear Strength of Sheeting Joints

Based on observations of many natural slopes it appears that failure of slopes along sheeting joints occurs predominantly by translational sliding of slabs of rock, often initiated by water pressure in the joint network. The problem is therefore relatively tractable to evaluate as it essentially only involves planar failure calculations rather than more complex wedge intersection displacements involving two or more joints. The common persistent nature of such joints means that the difficult judgmental issue of the contribution from true cohesion from rock bridges is of minor importance although cohesion might be a real factor for infilled and weathered zones within sheeting joints and for stepped situations where different sections of joint terminate against a pre-existing cross joint. Key factors that always need consideration are geometry (orientation and roughness at all scales), shear strength and the potential for ingress and development of adverse water pressure.

Sheeting joints characteristically are limited to shallow depths and therefore they are subject only to low confining stress. The fact that they tend in most domains to be rough, wavy and often persistent over considerable distances means that they are amenable to rational assessment of shear strength. It is generally agreed that the shear strength of persistent joints can be considered as derived from some "basic" frictional resistance offered by an effectively planar, natural joint plus the work done in overriding the roughness features on that joint. This is generally expressed by the following equation (after Patton 1966):

$$\tau = \sigma \tan(\phi_b^\circ + i^\circ)$$

where τ is shear strength, σ is normal stress, φ_b° is a basic friction angle for a planar joint and i° is a dilation angle that the center of gravity of the sliding slab follows during shear, i.e. the deviation from the direction that the shearing would have followed if the plane had been flat and sliding had occurred along the mean dip direction of the joint. Despite the apparent simplicity of the Patton equation, derivation of the parameters can be difficult, especially for judging the effective roughness angle.

BASIC FRICTION, Φ_B

Basic friction of natural joints can be measured by direct shear testing but tests need very careful setup, instrumentation and analysis if they are to make sense. A series of tests on different samples of a joint will often yield very wide scatter which is simply not interpretable without correcting for sample-specific dilation as described by Hencher and Richards (1989) and Hencher (1995). Dilation reflects work being done in overriding asperities. The dilation angle, measured during a shear test will vary especially according to the original roughness of the sample and the stress level. It is test-specific, will vary throughout a test and with direction of testing. It is not the same as the dilation angle, $i°$, which needs to be assessed at field scale, although the mechanics are the same. To avoid confusion the laboratory-scale dilation angle measured during a test is here designated, $\psi°$, whereas the field-scale dilation angle to be judged and allowed for in design is, $i°$, as defined by Patton 1966).

Typically, because of the complex nature of shearing, with damage being caused to some roughness asperities whist others are overridden, the dilation angle, $\psi°$, is difficult to predict for an irregularly rough sample although numerous efforts have been made to do so with some limited success (e.g. Archambault et al. 1999; Kulatilake et al. 1995). In practice, rather than trying to predict dilation which will be unique to each sample, stress level and testing direction, it is a parameter that needs to be measured carefully during direct shear tests so that corrections can be made to derive a normalized basic friction angle for use in design. Figure 13 shows the result from a well instrumented direct shear test on a rough interlocking joint. It can be seen that the peak measured ratio between shear stress and normal stress (τ/σ) is about 1.4 corresponding to a peak uncorrected friction angle, $\varphi_p = 54°$, given that $\tau = \sigma \tan(\varphi_p°)$. This peak strength however includes the effect of the upper block having to override the roughness as the joint dilates and work is done against the confining pressure. The dilation contribution is specific to this sample, this stage of the test, stress level and direction of shearing so cannot be taken as representative or even indicative of the roughness component of the joint, $i°$, at field scale. The measured dilation angle, $\psi°$ (which equals $\tan^{-1} dv/dh$ where dv is the vertical displacement increment over horizontal displacement increment, dh), and which varies throughout the test, can however be used to correct (normalize) the shear strength incrementally using the following equations:

$$\tau_\psi = (\tau \cos \psi - \sigma \sin \psi) \cos \psi$$
$$\sigma_\psi = (\sigma \cos \psi - \tau \sin \psi) \cos \psi$$

where τ_ψ and σ_ψ are the shear and normal stresses corrected for dilation caused by sample roughness. In practice, experience shows that for a system measuring to an accuracy of about ±0.005 mm, analysis over horizontal displacement increments of about 0.2 mm generally gives reasonably smooth dilation curves whilst retaining most of the detail of the test (Hencher 1995).

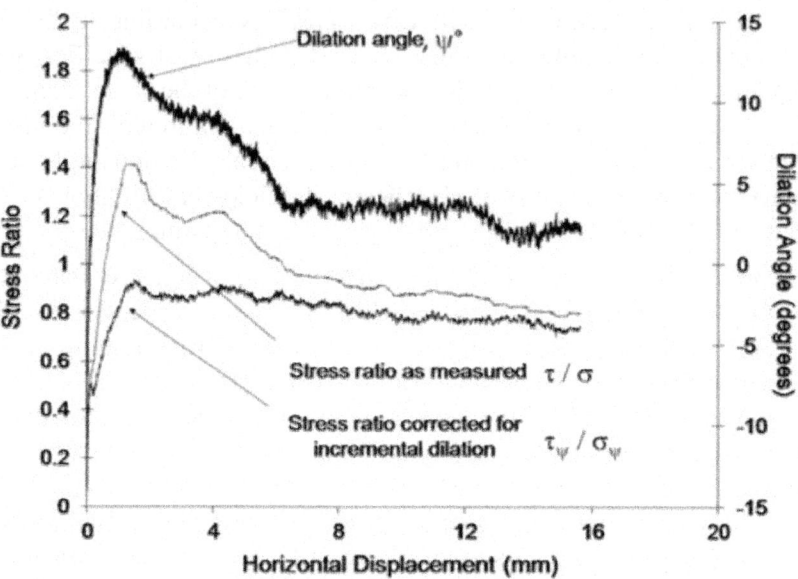

Figure 13: Typical test from single direct shear test run on a joint sample. Note how the measured strength strongly reflects the incremental dilation angle and how the corrected strength (lower line) is essentially constant once the influence of major, sample-specific roughness has been accounted for.

In Figure 13 the maximum dilation angle is about 15° and occurs slightly earlier than when peak shear strength is measured and then reduces throughout the rest of the test. The ratio between incrementally corrected shear and normal stresses (τ_ψ/σ_ψ) peaks at about 0.9 which means a dilation-corrected basic friction angle of about 42° (tan^{-1} 0.9). The friction angle then reduces gradually over a further 14 mm horizontal displacement to about 38° which reflects a reduced textural interlocking and ploughing component together with the production of gouge (Rabinowicz 1965; Scholtz 1990). Tests can be run multi-stage in which the same sample is used for tests at different confining stresses. These test stages generally give the same corrected friction despite the dilation angle changing with stress level and as damage is done to the surfaces. Tests must be properly documented however with photographs, sketches and profiles so that any variable data can be explained rationally (Hencher and Richards 1989). Generally it is found that tests on a series of samples from the same joint set (with similar surface mineralogy and textures) provide a reasonably well-defined dilation-corrected strength envelope as illustrated in Figure 14. That strength is frictional (obeys Amonton's laws) and comprises an adhesional component plus a non-dilational damage component that varies with textural and second order roughness (Hencher 1995).

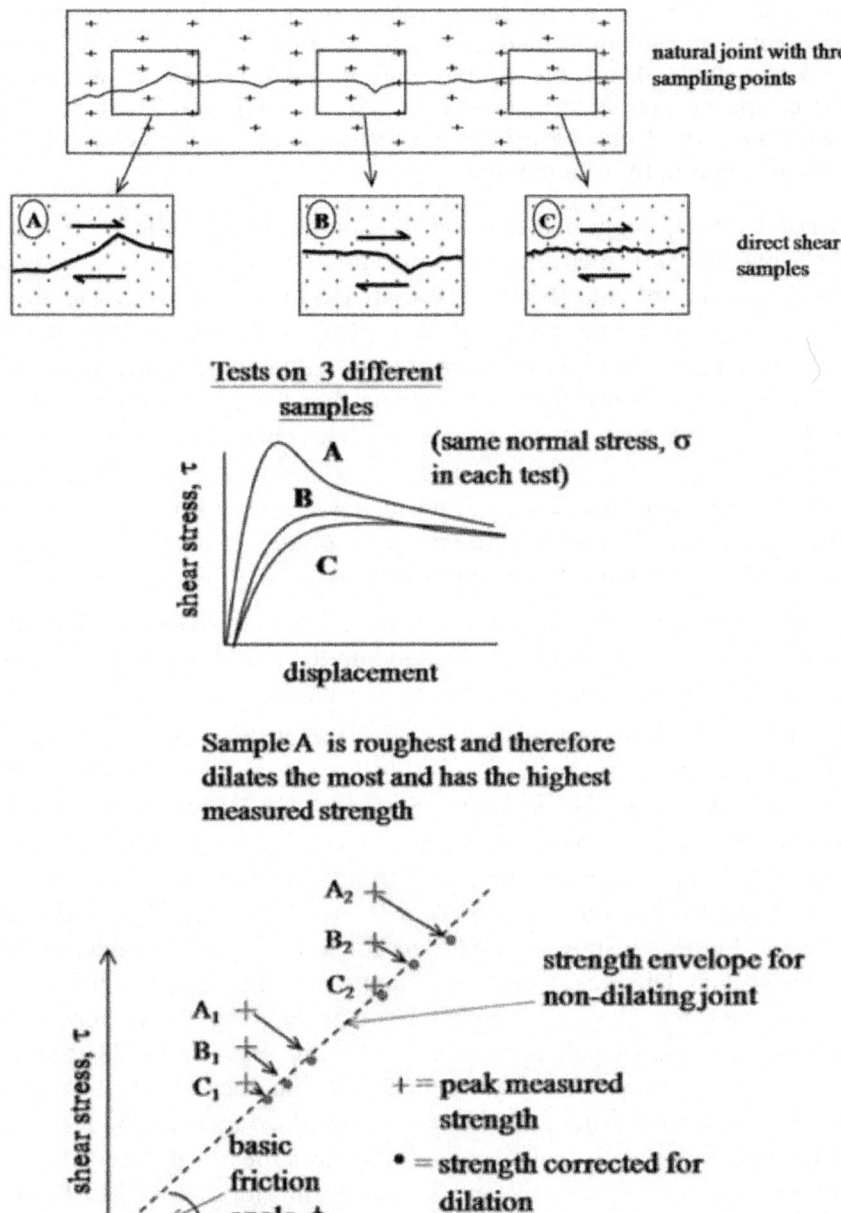

Figure 14: Series of direct shear tests carried out on three samples taken from different locations along the same joint. Different roughnesses will result in

different peak strengths for the same normal stress but dilation correction will reduce scatter considerably and give a basic friction for naturally textured, non-dilatant joints. The same samples can be tested repeatedly at increasing or decreasing normal stress stages. Peak strengths will vary with amount of damage and normal stress level but the corrected strength envelope for basic friction, φ_b, is usually well defined.

Barton (1990) suggested that the dilation-corrected basic friction angle might be partly scale-dependent as assumed for the asperity damage component in the Barton–Bandis model (Bandis et al. 1981) but further research using the same shear apparatus and modelling setup as Bandis (1980), but with better instrumentation, indicates that this is unlikely (Hencher et al. 1993; Papaliangas et al. 1994). Rather it appears that the dilation-corrected basic friction, once the effects of small-scale roughness have been corrected for, as described above, remains fairly constant and seemingly independent of the length of the sample. Scale effects do however need to be considered as a geometrical effect when deciding on the appropriate field-scale i° value to add to the dilation-corrected φ_b as discussed below.

This suggested procedure of first testing joints to determine a dilation-corrected basic friction angle and then adding the field-scale roughness angle component is best illustrated by some case examples.

In the early 1980s an extensive series of direct shear tests was conducted on sheeting joints samples taken from drill core as part of the North Point Rock Slope Study in Hong Kong. Samples included strong joints with quartz coating, joints coated with iron and manganese oxides and joints through highly decomposed, grade IV granite (Hencher and Richards 1982). Figure 15 shows dilation-corrected data for joints through grades II and III granite and Figure 16 shows similar data for joints through friable, grade IV rock. These data define essentially the same strength envelope with $\varphi_b \approx 40°$. Similar values for dilation-corrected basic friction angle are reported for other silicate rocks (Papaliangas et al. 1995) and Byerlee (1978) found the same strength envelope ($\tau = 0.85\sigma$) for a large number of direct shear tests on various rock types where dilation was constrained by using high confining stresses. Similar strengths have been reported as a mean value generally even for more weathered grades IV and V granite in Hong Kong (El-Ramly et al. 2005). Papaliangas et al. 1995 suggest, on the basis of these sort of results, that a friction value of about 40° for granite joints may mark a transition from dilational to purely frictional behavior and may relate to a change from brittle to ductile behavior within highly stressed asperities. Empirically it seems to be about the highest value for basic friction achievable for natural joints through many silicate rocks and applicable specifically to joints that are forced because of small-scale roughness to dilate, which includes most sheeting joints. That said, even higher dilation-corrected values can sometimes be measured for tightly interlocking, rough textured, tensile fractures through very strong rock, at least for several stages of testing (Hencher 1995, 2006). Conversely, it must be remembered that where joints are smoother so that they do not dilate during shear and where the surface texture is fine, polished or coated with low friction minerals such as chlorite, much lower basic friction angles can be measured for natural joints (Brand et al. 1983). This means that there is a marked variation between basic friction measured for artificially prepared (saw-cut and lapped) joints and for natural joints with different surface textures, as illustrated in Figure 17.

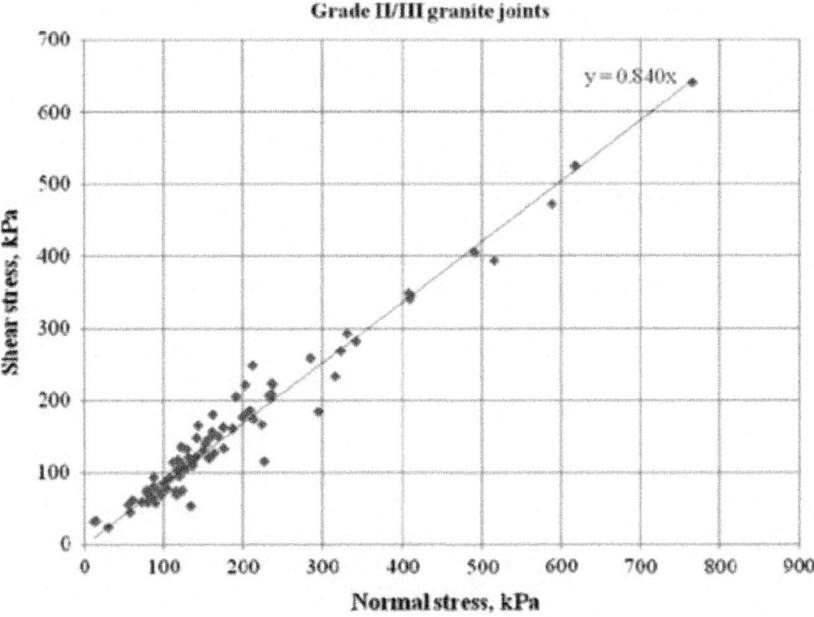

Figure 15: Dilation-corrected tests on sheeting joint samples through grades II and III rock, many samples iron- or manganese dioxide stained and with some patchy clay.

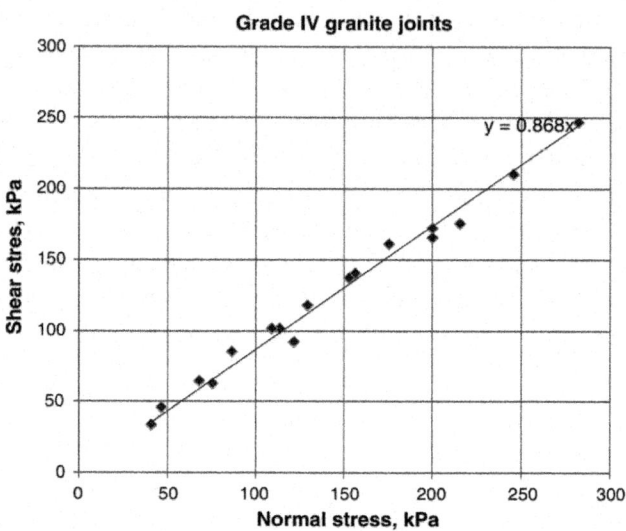

Figure 16: Dilation corrected data for grade IV material from sheeting joints, North Point, Hong Kong.

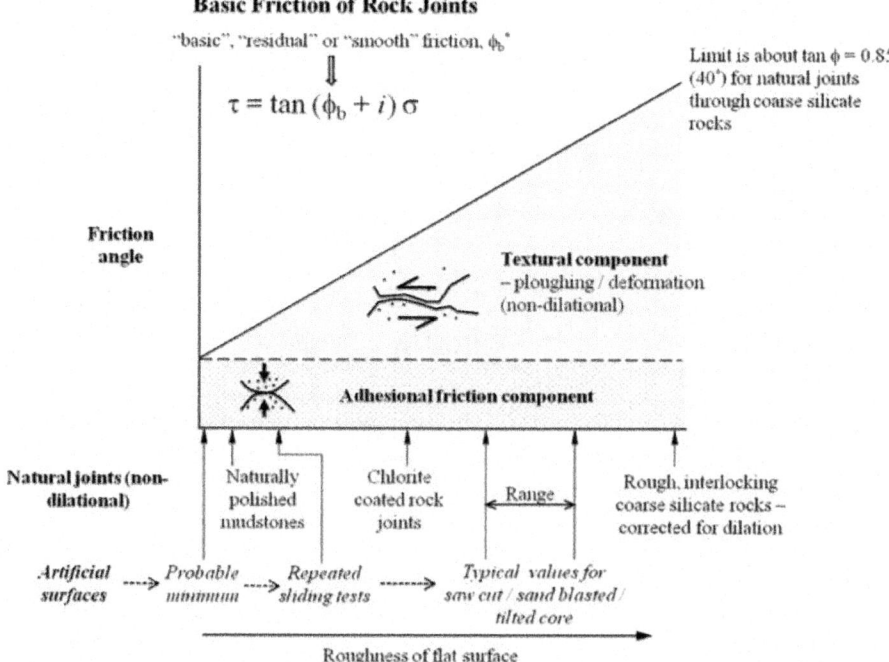

Figure 17: Components of basic friction for natural joints and artificial surfaces.

Roughness

Roughness at the field scale will be the controlling factor for the stability of most sheeting joints and for engineering design must be added to the basic friction φ_b of the effectively planar yet naturally surfaced and textured rock joint. Roughness is expressed as an anticipated dilation angle, $i°$, which accounts for the likely geometrical path for the sliding slab during failure (deviation from mean dip). There are two main tasks for the geotechnical engineer in analyzing the roughness component for a typical sheeting joint slab failure: firstly, to determine the actual geometry of the surface along the direction of likely sliding at all scales, and secondly, to judge which of those roughness features along the failure path will survive during shear and force the slab to deviate from the mean dip angle. This is the most difficult part of the shear strength assessment, not least because it is impossible to establish the detailed roughness of surfaces that are hidden in the rock mass. Considerable judgment is required and has to be balanced against the risk involved. Hack (1998) gives a good review of the options and the difficulties in exercising engineering judgment are discussed in an insightful way by Baecher and Christian (2003).

In practice, the best way of characterizing roughness is by measurement on a grid pattern in the way originally described by Fecker and Rengers (1971) and adopted in the ISRM recommended methods (1978) although spatial variability may be an important issue for sheeting joints; the important first order roughness represented by major wave features may vary considerably from one area to another. At one location a slab might be prevented from sliding by a wave in the joint surface causing a reduction in the effective down-dip angle along the sliding direction; elsewhere, a slab of perhaps several meters length may have a dip angle steeper than the mean angle for the joint as a whole because it sits on the down slope section of one of the major waves as illustrated by a case example later. For the Tuen Mun highway stability studies (Carter et al.2002) numerous true scale survey profiles down the plunge line of exposed sheet structures were collected using EDM surveying techniques, with abseil approaches being utilized to achieve profiles down even the steepest of the slope gradients. These detailed profiles of the exposed surfaces were assumed to be representative of adjacent hidden joints that were candidates for potential failure. The level of roughness/asperity surface detail that is attainable by this sort of survey profiling is illustrated for an extensive discontinuity in Figure 18.

Figure 18: Roughness survey under way using plates of different diameter on a bedding plane discontinuity in shaley limestone (Skipton, Yorkshire). At the small scale (hammer) the joint is rough but at larger scale the joint is effectively flat.

Defining the scale at which roughness will force dilation during sliding rather than being sheared through requires considerable judgment. Some assistance is provided by Schneider (1976) and by Goodman (1980) who indicate that for typical rough sheeting joint surfaces, where slabs are free to rotate during shear, as the length of the slab increases (at field scale) the dilation angle controlling lifting of the center of gravity of the upper block will reduce. As noted earlier, sheeting joints are often wavy and major waves, where opposing the shearing direction, can almost always be relied upon to cause dilation at field scale from the mean dip of the overall sheeting joint plane, especially at the low stress levels appropriate for most sheeting joints despite the obvious stress concentrations at overriding contacts. Simple geometry shows that for wave amplitude of 1 m over a wavelength of 20 m the minimum dilation angle would be about 6°; over a wavelengths of 10 m, 11°; and over 6 m, 18°. This is an example of where geometrical scale effects operate and must be taken into account.

The characterization of the geometry of sheeting joints in the field is described by Richards and Cowland (1982). Specifically for the North Point Study in Hong Kong it was judged that relatively small base length asperities in grade II and III rock would survive the field stresses based on the observation that little damage occurred to even smaller asperities at stress levels higher than in the field during direct shear tests (e.g. Figure 19). They therefore concluded that the angle of deviation from mean dip of joints could be relied upon to be 16° for long sheet joints with lengths of >20 m, based on the geometry of frequently occurring asperities (they judged that some second order asperity features could be relied upon as well as first order roughnesses). Richards and Cowland (1982) appreciated that the dip of sheeting joints can suddenly steepen on one side of a wave thereby increasing the local dip for a relatively short (few metres) length slab of rock as illustrated by the Hui Ming Street landslide later in this paper. In summary, the issue for differentiating the contribution of small- and large-scale asperities boils down to carrying out appropriate rock characterization. The problem cannot be finessed by improved analytical methodology. There is no substitution to careful engineering geological inspection, investigation and judgment based on experience of similar joints and geological settings and an appreciation of the fundamental mechanics controlling the potential failure.

Figure. 19: Sample V13, 8.7 m (North Point Study, Hong Kong) following 5 stage, repeated direct shear test up to normal stress 285 kPa equivalent of more than 10 m confining stress. Note the localised nature of damage (white areas) on the weathered, stepped surface coated in brown iron and black manganese dioxide. Note that main step feature has survived intact (and was responsible for dilating the joint) at that stress level.

Infilled Joints

Richards and Cowland (1982) suggest that where there is a thick band of weathered rock along a joint (say grades IV and V) as shown in Figure 5, zero dilation should be allowed when assessing stability. Where the joint is infilled with a mixture of weathered rock and rock fragments however the Hoek–Brown strength criterion might be used to provide some estimate of strength without laboratory testing (Carter et al. 2002) although Brown (2008) cautions against applying the original criterion outside the original data set and expresses specific concern for application for rocks with uniaxial compressive strength (UCS) below about 30 or 40 MPa. Carter et al. (2008) and Carvalho et al. (2007) discuss a modified Hoek–Brown criterion for low strength rocks that may be more applicable for such application.

As discussed earlier, incremental movement of sheeting joints may take place over many years before final slab detachment and, following each movement, sediment may be washed in to accumulate in hollows on the joint (Figure 20). The presence of washed-in sediment may indicate that the rock mass has moved but this is not always the case as illustrated by a case at Kwun Tong Road, Hong Kong, discussed later, where the observed sediments were deposited in an erosional pipe along a sheeting joint rather than in a void opened up by dilation. The presence of in-washed sediment in a joint might cause alarm during ground investigation (clay infill having relatively low shear strength) but in many cases such sediments are patchy in occurrence and confined to local down warps on a partially dilated joint. The sediment is probably playing little or no part in decreasing frictional resistance which is controlled by contact between rock asperities.

Figure. 20: Hard, slickensided kaolinite infill from downwarp on failed sheeting joint (Hencher 1983). Pencil for scale.

When the infill becomes of such thickness that rock wall contact is no longer to be relied upon then of course the infill strength itself needs to be assessed for design. The infill will also affect hydraulic conductivity properties of such infilled joints and particularly will affect pore pressure dissipation rates, potentially leading to an adverse stability state due to the potential for a lengthened period of reduced effective stress following a rain storm event. Experience from the piezometric monitoring of "infilled" and of "clean" sheeting joints at Tuen Mun (Carter et al. 2002) showed that although similar maximum pore pressure spike levels were recorded for both joint types, it took days instead of hours for the dissipation of the excess head in the filled joints compared with the clean joints.

Estimating Shear Strength Using Empirical Methods

Because of the inherent difficulties, need for quality equipment and expertise required for measuring shear strength of rock joints, various empirical criteria have been proposed for estimating shear strength based on index tests and idealized joint shapes. The most widely used strength criterion is that proposed by Barton (1973). This takes the "basic friction" measured for saw-cut or other artificially prepared planar surfaces then adding in a component to account for roughness adjusted for the strength of the rock asperities and for scale. Details are given in many text books including Brady and Brown (1985) and Wyllie and Mah (2004). The advantages of this criterion are its apparent ease of use and application in numerical modelling but there are difficulties in determining each of the various parameters. Basic friction is taken to be a lower bound component with a "limiting value" of 28.5–31.5° (Barton and Bandis 1990) but the friction measured for a saw-cut surface is not necessarily a lower bound either for natural or artificial joints (see Figure 17). Hencher (1976) for example reports the sliding angle reducing from about 32° to only 12° for saw-cut surfaces of Darleydale Sandstone after about 4 m in tilt tests with continual removal of rock flour between test runs. Furthermore, considerable variability is sometimes reported from tests carried out on artificially prepared surfaces. Stimpson (1981) measured values ranging from 24° to 38° using limestone core pieces in sliding tests. Tests reported by the Norwegian Geotechnical Institute (NGI) for the Åknes landslide investigation gave values ranging from 21° to 36.4° for tilt tests on saw-cut joints with about 73% of data between 25° and 30.2° (Kveldsvik et al. 2008). Nicholson (1994) reports a variation in 12.5° for tests carried out on carefully prepared saw-cut, lapped surfaces of Berea sandstone, all suggesting that the recommendation of some lower "limiting value" of 28.5 to 31.5° may not be universally applicable. There is also some confusion in the literature regarding application of some of the Barton early equations as to whether φ_b (which is stipulated as sawn surface value determinations) or φ_r (residual values from multi-reversal shear box testing) is the appropriate parameter for application in the equation. In the authors' opinion is also extremely unwise to rely on the widely publicised

Schmidt Hammer relationships proposed between residual strength and base friction angle as a means for sorting out the correct value for shear strength determination.

The contribution to shear strength from roughness for small-scale roughness can be measured or estimated from standard shape profiles, but this can be difficult in practice and varies according to shearing direction and with scale, requiring appropriate judgement for its effective application. Beer et al. (2002) carried out an online survey of people's estimates of joint roughness coefficient (JRC) for three randomly selected joints. Considerable scatter was reported and for one of the three joints a possibly bi-modal distribution of estimates was determined with the two centers of population at 8.9 and 17.9, perhaps reflecting different individual's perception of controlling roughness scale. Like any other stochastic parameter, considerable difficulties can occur when representing joint roughness with a single value JRC estimate, as clearly demonstrated by determinations for the Åknes landslide by workers from NGI and MIT (Kveldsvik et al. 2008) where JRC measured for foliation joints at a 0.25-m scale ranged from 2.5 to 20 with a mean of 10.6. At a 1-m scale, JRC estimates covered the full possible range (from 0 to 20) with a mean of about 8 and standard deviation of ~4. The range of calculated factor of safety for this range of JRC was from about 0.8–2.0 taking all other parameters at their mean values. As is obvious, considerable judgment is still needed in application of such empirical procedures so that overall estimates for joint surface strength can be considered realistic. Furthermore, once the second order roughness contribution has been decided upon, then an additional roughness angle, $i°$, still needs to be determined and added, to account for larger scale roughness not sampled in the JRC assessment (Barton 1990). An important point that arises from this review of empirical strength criteria for estimating field strength of rock joints is that it needs to be emphasized that the correct base-line parameters must be utilized within the equations whatever approach is adopted. It is a prevalent misconception in the literature (e.g. Simons et al.2001) that dilation-corrected data from direct shear tests on natural joints can be used interchangeably in empirical equations. This is incorrect because the dilation-corrected strength already includes a frictional component contributed from textural and roughness damage (part equivalent of JRC) and its substitution for the saw-cut or residual φ_b of Barton could lead to overestimations of field-scale strength by maybe 10° in many cases.

CASE EXAMPLES OF LANDSLIDES INVOLVING SHEETING JOINTS

A number of landslides involving sliding on sheeting joints have been studied in some detail in Hong Kong and provide some insight into operative shear strength and mechanisms of failure.

Sau Mau Ping Road, Hong Kong, Early 1970s

An example of an in-depth study of slope stability governed by potential sliding on sheeting joint is presented by Hoek (2009) and Wyllie and Mah (2004). Figure 21a shows a section of the slope as it is today, little changed from the slope photographed in the early 1970s, with an extensive section of exposed sheeting joint following a failure during blasting to construct the road. It was anticipated that the exposed sheeting joint would extend through the adjacent 60-m-high slope with an overall angle of 50° and individual batters 20 m high and inclined at 70°. The slope that was of concern has now been cut back as illustrated in Figure 21b (compare to Hoek's Figure 4). Hoek goes through a reasoned series of sensitivity calculations based on various assumptions, culminating in the decision to cut back rather than drain or reinforce the slope. The interpretation at the time was that the additional strength offered by dilation in overcoming roughness could be expressed as apparent cohesion. If these analyses and calculations were repeated today probably a slightly different approach would be taken in the way that shear strength was dealt with and consequently on the measures adopted.

Figure. 21: **a** View of sheeting joint (2007) pictured in Hoek (2009); **b** picture of the slope analysed and then cut back as discussed in detail in Hoek (2009).

In particular a cohesive component of strength is insensitive to water pressure assumptions, whereas if strength is expressed as friction plus dilation angle, both of these are dependent on effective stress and a different answer would ensue. Apparent cohesion is clearly a good concept for jointed rock masses (e.g. Brown 2008) but not for the shear strength of persistent rock joints. The result and conclusions might still be the same (cutting back the slope to the dip angle of the sheeting joint is certainly a pragmatic solution) but the assessment approach might well now be different.

Hui Ming Street 2000 and 1993

In 2000 a block of rock with volume of 15 m³ fell close to a playground after sliding on a sheeting joint. The source of the rock fall is shown in Figure 22. A nearby previous failure of a 20-m-wide section of slope in 1993 was re-examined as part of the study into the 2000 landslide Halcrow China Ltd (2002a). The basal surface of the 2000 rockfall dipped out of the slope at about 35°–38°. Using a 420-mm-diameter plate on a 200-mm grid across the failure surface a dilation angle, i° of 8°, was determined and for an 80-mm plate, between 12° and 14° which is consistent with measurements on other sheeting joints in Hong Kong. The section of sheeting joint below the failed block that would have been exposed prior to failure, dipped more shallowly at less than 30°. This case illustrates that significant block failures can occur on unexposed steep sections of sheeting joint. Back analysis showed that for an effective friction angle (φ_b + i) of 43° the factor of safety was 1.2 under dry conditions and reduced to 1.0 with water pressure of about 10 kPa. The nearby, much larger 1993 failure surface was investigated by cutting trial trenches through shotcrete that had been placed over the landslide scar. The mean dip of the lower part of the failure surface was only about 22° but steepened to about 45° over the rear half of the failure surface. Roughness values (deviation from mean dip) were measured as 16° using an 80-mm plate and 12° for a 210-mm plate. Much of the basal surface was however coated with completely and highly decomposed granite which would have reduced the effective dilation angle.

Figure. 22: Location of rock fall at Hui Min Street, 2000.

Above Leung King Estate: 2000

The progressive nature of failures associated with sheeting joints is illustrated well by a landslide that occurred above Leung King Estate, Hong Kong (Halcrow China Limited 2002b; Hencher 2006). Features of this landslide are illustrated in Figure 23. Surviving remnants of the rock mass above the main sheeting joint along which detachment finally took place showed signs of long-term movement, growth of fractures and sediment infill prior to failure. The stages prior to failure are shown schematically in Figure 24.

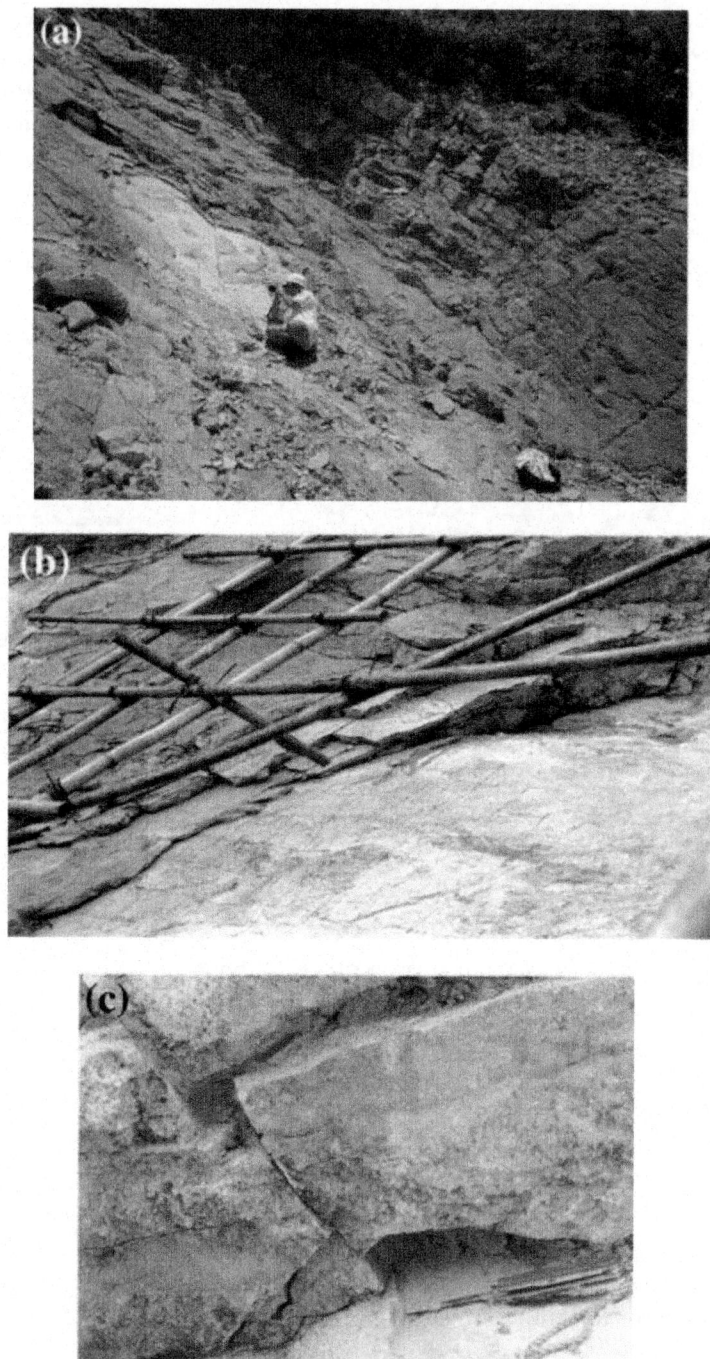

Figure. 23: Landslide above Leung King Estate, Hong Kong. **a** Below the detachment

surface the rock is light colored and joints are rare; above, the rock is fractured and discolored. **b** Lower wavy surface with grid for roughness measurements using plates. **c** Side view with evidence of long-term displacement including dilation and deposition of sediments in voids.

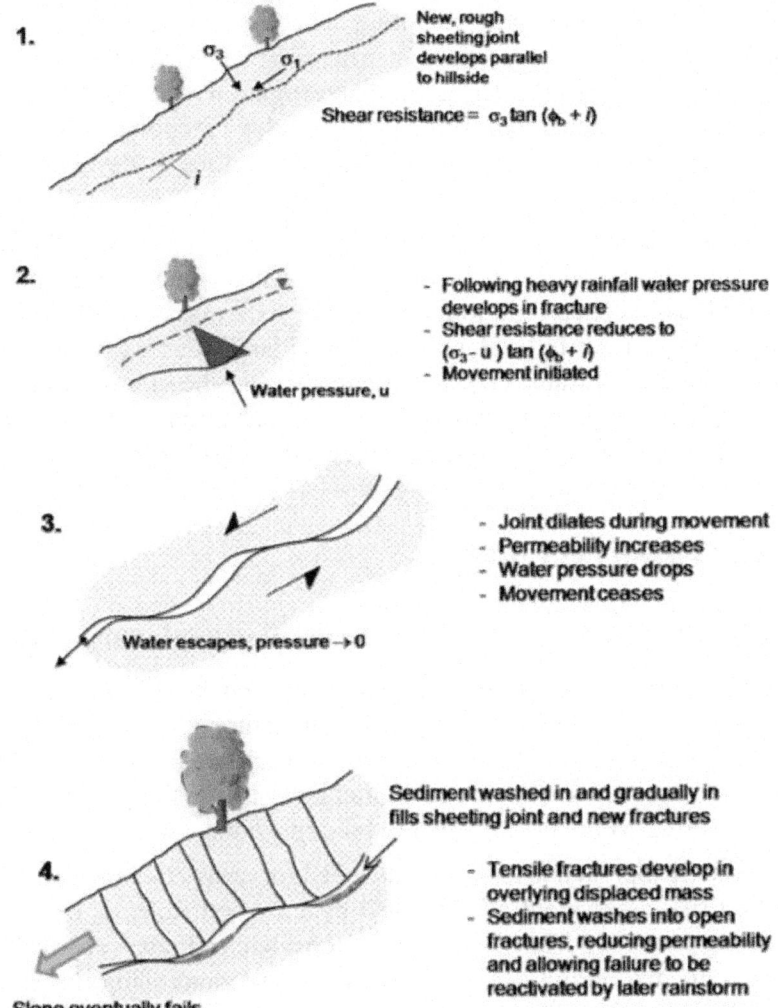

1. σ_3 σ_1

New, rough sheeting joint develops parallel to hillside

Shear resistance $= \sigma_3 \tan(\phi_b + i)$

i

2.

Water pressure, u

- Following heavy rainfall water pressure develops in fracture
- Shear resistance reduces to $(\sigma_3 - u)\tan(\phi_b + i)$
- Movement initiated

3.

Water escapes, pressure $\rightarrow 0$

- Joint dilates during movement
- Permeability increases
- Water pressure drops
- Movement ceases

4.

Sediment washed in and gradually in fills sheeting joint and new fractures

- Tensile fractures develop in overlying displaced mass
- Sediment washes into open fractures, reducing permeability and allowing failure to be reactivated by later rainstorm

Slope eventually fails

Figure. 24: Schematic representation of history of Leung King landslide prior to detachment.

Lessons from Landslide Case Studies

These case examples of landslides involving sliding on sheeting joints have provided some useful insights into the nature and characteristics of such failures. In particular, the failure above Leung King Estate gave considerable evidence of long-term deterioration involving intermittent movements by sliding along the joint along which detachment eventually occurred. Such deterioration with sediment infill and natural pipe systems may be taken as indications that the slope may be failing. The importance of the development of cleft water pressure is evident in triggering most sheeting joint failures investigated in Hong Kong. The ground investigation reported by Richards and Cowland (1986) which demonstrated the complex reaction of water pressures in joints to rainstorm events indicates the difficulties in designing drainage measures to prevent failure.

The case studies demonstrate the difficulties in extrapolating the geometry of sheeting joints into the rock mass from measurements in exposures. In particular local increases in dip hidden in the rock mass can allow significant and unpredicted rock falls. Back analysis of landslides adopting reasonable estimates for active water pressures confirms that the current approaches to assessing shear strength based on dilation-corrected basic friction plus an i° value judged from roughness measurements on a grid basis at field scale can provide realistic parameters for design use.

ENGINEERING WORKS

Assessing Risk and the Need for Preventive Engineering Measures

Slopes in sheeting joint terrain often appear extremely threatening because of the persistent, daylighting and steeply dipping nature of the joints. The fact that such steeply dipping joints are associated with failures at all scales from small rock falls to major translational movements has, over the years, necessitated that engineering works be implemented to reduce the risks.

A modern approach to assessing the need for preventive measures is to use quantified risk assessment as described by Pine and Roberds (2005) for the widening of the Tuen Mun Highway in Hong Kong (Figure 1b). This project involved remediation and stabilisation of several sections of high cut and natural slopes dominated by potential sheeting joint failures and by the potential for failure of rock blocks and boulders bouncing down exposed sheeting joints to impact the road below. Design of the slope cut backs and stabilisation measures was based on a combination of reliability criteria and conventional Hong Kong standard factor of safety design targets aimed at achieving an ALARP (as low as reasonably possible) risk target which, in actuarial terms, translated to less than 0.01 fatalities per year per 500 m section of the slopes under remediation.

General Considerations

Remediation of sheeting joint-controlled stability hazards on high rock slopes is often not trivial and implementation of the works can itself increase the risk levels albeit temporarily. Factors that will influence the decision on which measures to implement include the specific nature of the hazards, topographic and access constraints, locations of the facilities at risk, cost and timing. The risks associated with carrying out works next to active roads both to road users and to construction workers themselves and how to mitigate these are addressed in some detail in Geotechnical Engineering Office (2000a) and Halcrow China Limited (2002c). Pre-contract stabilization works will often be needed to allow initial site access and preparation. Preventive measures such as rock bolting may be carried out at an early stage to assist in the safe working of the site and designed to form part of the permanent works. Options for the use of protective barriers and catch nets to minimize disruption to traffic during the works also need to be addressed, as do contractual controls and alternatives for supervision of the works. The use of a risk register, as piloted for tunnels (Brown 1999), with clear identification of particular risks and responsible parties, helps to ensure that all hazards and consequences are adequately dealt with during construction. Decision analysis is now widely applied at an early stage to assess whether to mitigate slope hazards (e.g. by rockfall catch nets) or to remediate/resolve the problem by excavation and/ or support approaches. If construction of intrusive engineering measures to stabilize hazards might be unduly risky, then passive protection can be adopted instead. A hybrid solution is often the most pragmatic solution for extensive, difficult slopes such as at Tuen Mun Road where some sections were stabilized by anchors and buttresses and other sections were protected by nets and other measures (Carter et al. 2002; Pine and Roberds 2005).

ENGINEERING OPTIONS

Some of the options for improving the stability of slopes are listed in Figure 25 and illustrated in Figure 26. These can be split into passive options that either deal with the possible failure by controlling surface deterioration at source or installing preventative reinforcement to increase local factors of safety, or adding walls or buttresses to restrain detached debris before it causes injury or damage and active measures that enhance overall factors of safety of larger sections of slope by major engineering works including cut backs or buttresses or heavy tie-back cabling.

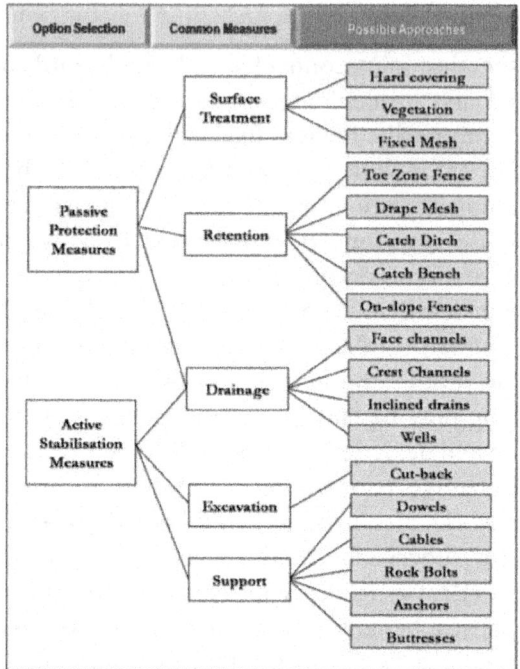

Figure. 25: Engineering options for stabilizing slopes in sheeting joint terrain.

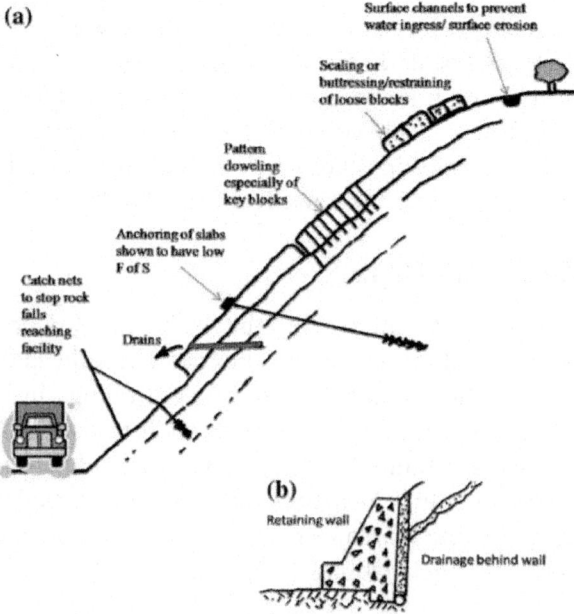

Figure. 26: **a** Main engineering options for stabilizing slopes in sheeting joint terrain; **b** retaining wall used to buttress large section of slope.

Surface Treatment

Many risks can be mitigated through surface treatment to stabilize or remove relatively small blocks of rock. There is a temptation to use hard slope treatments like chunam (old HK remedy) and shotcrete to constrain loose blocks at the slope surface but such measures if not properly designed can restrict drainage from the slope, hide the geological situation from future investigators and can themselves cause a hazard as the shotcrete deteriorates allowing large slabs of shotcrete to detach and impact whoever is unfortunate enough to be below (Figure 27). Furthermore, shotcrete is increasingly an unacceptable solution for aesthetic reasons and there is a push towards landscaping high, visual slopes where measures can be justified from an engineering sense (Geotechnical Engineering Office 2000b). In this context it is to be noted that most bio-engineering solutions will not work for high-risk slopes in that they cannot be relied upon in the long-term and root growth can lead to blocks becoming loosened and detached. Where individual rock fall sources are identified, these can be scaled off, reinforced by dowels, bolts, cables or dentition and/or netted where the rock is in a closely jointed state. Removing large blocks can be difficult because of the inherent risks associated with breakage techniques including blasting and chemical splitting which can dislodge blocks unexpectedly. Care must be taken to protect the public and workers during such operations. The most difficult zones to deal with are those with poor access. Implementing passive or active protection approaches needs to start from safe ground and move progressively into the areas of more hazardous stability.

Figure. 27: Use of shotcrete on broken rock face together with anchorages (nails). Disadvantages include damming water, hiding what is going on and the shotcrete itself will form a hazard as it deteriorates.

Rockfall trajectory analysis using widely available software allows prediction of energy requirements and likely bounce heights and run-out damage zone extent. Where energy considerations allow, toe-zone protection measures, catch benches, catch ditches, and toe fences provide the earliest viable mitigation approach without requiring access on the slope.

Surface drainage is a very important consideration for all slopes but particularly for slopes comprising part rock (with very high runoff) and soil sections which might be eroded and undermined from high surface flow concentration.

Mesh Drapes

Where slope heights are significant and ramp or bench approach is difficult, mitigating hazards can be problematic even using rope access techniques because face stability may be too unstable to even allow rock climbing personnel onto the face. Under such conditions surface mesh draping may allow some effective protection to be achieved preventing ski jump-style bouncing of rock progressively down slope (Carter et al. 2002). Application of drape mesh (varying from chain-link, triple twist, hex-mesh to ring-net in increasing order of energy capacity) can be effected by a variety of techniques ranging from climber controlled unrolling of the mesh to helicopter access placement. Typically, crest restraint is provided by dowels or tie-back anchors usually cabled back some distance from the crest zone to provide a safe anchorage.

Fences, Catch Nets and Barriers

Where there is the potential for repeated small-scale detachments impacting a highway, then catch nets or diversion/stopping barriers can be the solution as discussed with reference to reducing risk by Pine and Roberds (2005). Such catch nets or fences can be positioned on-slope as illustrated in Figure 28 (from Carter et al. 2002) or in the toe zone of the slope depending on energy requirements and site restrictions. Where energies computed from rockfall analyses are too extreme for toe-zone protection alone to maintain risk levels below prescribed criteria for highway users, on-slope energy protection fences become a necessity to reduce total energy impact at road level. This was the approach adopted at Tuen Mun Road in Hong Kong for sections of the slopes which were to remain in place and where sheeting joint geometries were considered hazardous enough to allow potential release of blocks of sizes that could not be stopped by toe-zone fencing alone. The photograph in Figure 28 shows an on-slope 3,000 kJ fence designed and installed above the highway to catch rockfall blocks from the 100 m of slope upslope of the fence.

This fence is located about 80 m vertically above the main carriageway of the highway, where the main toe-zone fence and catch ditch are located.

Figure. 28: Catch net to stop rock falls, above Tuen Mun Highway, Hong Kong.

Drainage

Drainage can be very effective in preventing the development of adverse water pressures, but there is a need to target subsurface flow channels many of which will be shallow and ephemeral. The paths may be tortuous and hard to identify and drainage measures can therefore be rather hit or miss (Hencher 2010). Regular patterns of long horizontal drain holes can be very effective, but it must never be expected that all drains will yield water flows and the effectiveness of individual drains can change with time as subsurface flow paths migrate. With exposed sheeting joints forming ledges on a slope, care must be taken that the step zones are not shotcreted otherwise free drainage may be impeded and water might dam up behind the shotcrete. If the exposed joint is weathered the weak material may back-sap and possibly pipe leading to destabilisation, partially caused by lack of free drainage. This can be rectified by installing closely spaced horizontal drains with geotextile filter fabric sleeves so as to prevent blocking together with protection of the weathered material. No-fines concrete whilst appearing to be suitable to protect weathered zones often ends up with lower permeability than designed and should not be relied upon without some additional drainage measures.

Reinforcement

The factor of safety against slab sliding can be improved by a variety of options. For sheeting joints specifically, provided there has not been previous movement, the rough interlocking nature of these tension fractures provides considerable shear strength (where not severely weathered) and this needs to be accounted for in design in order to avoid over-conservatism. If the joint can be prevented from sliding by reinforcing at strategic locations then full advantage can be taken of the considerable natural frictional resistance. Active stabilisation of blocks is possible if they are of relatively small size and access is feasible either by rope access techniques down the slope, using "spyder" drills or even better if tracks can be constructed, using more conventional drilling equipment. Depending on configuration, rock blocks may be stabilised by dowelled concrete buttressing (to provide direct support to a well-defined potential release block), through various forms of tie-down and/or overturning control tie-back reinforcement, comprising deep sub-vertical dowelling. Sub-horizontal cable anchors can be used if capacities larger than about 20 tonnes per reinforcement member are required. Often the most significant reinforcement is needed where extensive sheeting joint zones define slabs of large proportions. In such cases, the preferred method in Hong Kong is to use passive dowel designs rather than tensioned bolting for necessary shear constraint. This is because it is considered that active reinforcement members are more subject to corrosion damage and that passive dowels allow both mobilisation of a normal force (due to the restraint provided by the full column bond against asperity ride during shear), plus active shear restraint provided by the steel of the dowels resisting block slide mobilisation (Spang and Egger 1990).

The Geotechnical Engineering Office in Hong Kong has published some guidelines on prescriptive measures for rock slopes and in particular gives guidance on rock dowelling for rock blocks with volume less than 5 m³(Yu et al. 2005). In essence, it is advised to use pattern dowels with one dowel per m³ of rock to be supported with minimum and maximum lengths of 3 and 6 m, respectively and where the potential sliding plane dips at less than 60°. The dowels are to be installed at right angles to the potential sliding plane, with the key intention to allow the dowels to act in shear, whilst also enhancing the normal restraint due to asperity ride during sliding. In practice dowels frequently need to be used in more variable orientations. It is often difficult to identify the thickness and volume of a given block requiring support and therefore dowel patterns frequently are based on some assessment of cross-joint spacing. Along the Tuen Mun Highway, typical support layouts were adopted based on field mapping of cross-joint spacing and orientation with respect to the sheeting joint geometry and inferred direction of sliding. The design used 40-mm dowels at 5-m spacing, based on analysis using the approach of Spang and Egger (1990) for definition of shear resistance. In areas of closer cross joints, 25-mm dowels were used at 2-m spacing, split spaced between the wider pattern bolting layouts (Pine and Roberds 2005). Field placement of reinforcement was however always double checked against natural disposition of features and decisions made by the engineer in the field for additional spot bolting or dowelling as required and illustrated in Figure 29.

Figure. 29: Spot bolting of sheeting joint slabs.

Buttressing

Where earlier failure or cutting exposes sheeting joints or detached blocks resting on a sheet structure then buttressing using a reinforced concrete wall or a raked structural prop may be an economic option and offer some certainty of solving potential sliding or toppling instability problems. One example is shown in Figure 30. Much of the slope illustrated has sheeting joints running through it. The slope failed progressively over a period of 2 years and the eventual solution was to fill the failed zone with a very large concrete buttress wall.

Figure. 30: Tsing Yi Island, Hong Kong. 500 m³ rock fall from sheeting joints: **a** debris cleared off and showing the extent of problem with extensive sheeting joints through the hillside; **b** the solution: a huge concrete buttress.

Figure 31 is a sketch of a large cut slope in Kwun Tong, Hong Kong, which was the location for numerous significant failures along sheeting joints over several years. The slope was investigated by subsurface boreholes and by face mapping and sections of the slope were designed to be re-profiled, soil nailed, dowelled and buttressed as appropriate. During construction however an additional sheeting joint was exposed unexpectedly and found to be partly open and partly infilled with stream sediments (Figure 32). It was established by careful mapping and matching geological features such as mineral veins across the joint that, despite the voids and sediment infill, the joint had not been displaced but simply eroded internally and locally. It was decided that the peak interlocking strength could still be relied upon and the solution adopted was to restrict any translational movement. Accordingly, a dowelled buttress wall was constructed to infill the area below the potential sliding slab with careful attention to ensure that the underground stream could continue to flow without restriction.

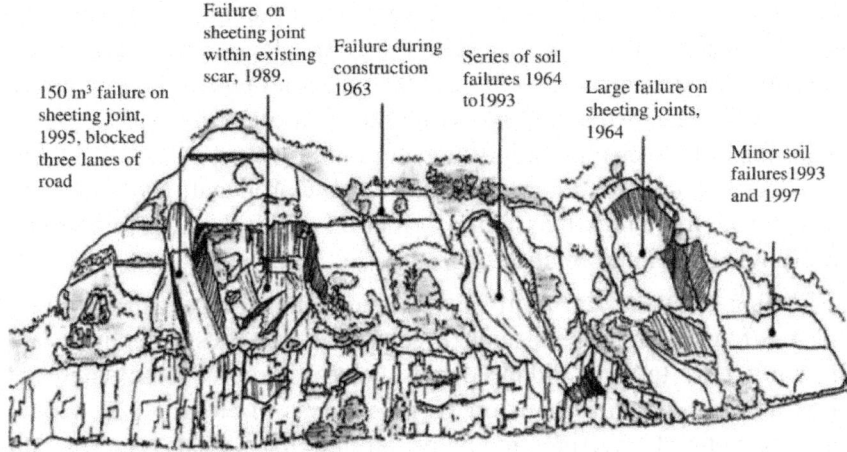

Figure. 31: Sketch of large section of rock slope along Kwun Tong Road (slopes up to 30 m high) with history of failures dominated by sliding on sheeting joints and selected for engineering upgrading work.

Figure. 32: Landslide preventive works underway at Kwun Tong, Hong Kong.

Similarly on the slopes above Tuen Mun Highway many combination buttresses were employed with parts of blocks dowelled and part buttressed as access and local geometry dictated.

CONCLUSIONS

Sheeting joints develop due to topographical or residual tectonic stresses close to the Earth's surface. Those that develop in natural slopes are often of adverse geometry with respect to natural hillside slopes and as such, may predispose the slope to repeated failure. Sheeting joint terrain often comprises a series of simple slabs resting on one another and often these are geologically young. Many other sheeting joints are very old however, as evidenced by their association with deep weathering profiles, by the propagation of other fracture systems through the rock mass after formation of the sheeting joints and from geomorphological interpretation.

Sheeting joints appear to always have originated by tensile opening and as such often occur as persistent, mechanical fractures extending laterally over many tens to even hundreds of metres. Detailed assessment of the configuration of sheeting slabs on various slopes in Hong Kong, Korea and other well-defined sheeting geologies around the world, suggests that in general, remnant slabs sitting on persistent sheet structures owe their stability more to the roughness and undulating/wavy character of the sheet structure (and associated dilation) rather than to rock bridges (and true cohesion) which is commonly the case for other types of joints. Definition of controlling shear strength is thus amenable to evaluation either through a testing programme combined with field measurement and assessment of roughness and analysis of the way that roughness will cause dilation or by employing empirical methods. Both approaches require considerable judgement.

Rock slope failure mechanisms based mainly on pseudo-statistical analysis of defect data should not be the sole basis of defining ground models. More intelligent analysis of the data is required and in the case of sheeting joints, recognition of lateral variation in orientation, roughness and degree of weathering and openness. The possibility for composite landslides, partly involving rough joints and partly through more weathered sections should be recognised. It is clear that sheeting joint failures are often associated with the development of cleft water pressures and that failure may be incremental over long periods and many storm events.

Landslide preventive works often necessitate reinforcement, drainage and rockfall protection (such as fences, catch nets and ditches). On steep hill slopes where detached or partially detached sheeting structures exist, buttresses and/ or anchor blocks have application for preventing initial movement that would otherwise lead to progressive deterioration. Preventing initial movement will optimise the contribution from peak shear strength. Seepage points on faces can help to identify the likely routes for channel flow which should be targeted with raking drains.

ACKNOWLEDGMENTS

This research was supported in part by a grant (NEMA-06-NH-05) from the Natural Hazard Mitigation Research Group, National Emergency Management Agency (NEMA), Ministry of Public Administration and Security, Korea.

REFERENCES

1.   Archambault G, Verreault N, Riss J, Gentier S (1999) Revisiting Fecker–Rengers–Barton's methods to characterize joint shear dilatancy. In: Amadei B, Kranz RL, Scott GA, Smeallie PH (eds) Rock mechanics for industry. Balkema, Rotterdam, pp 423–430

2.   Baecher GB, Christian JT (2003) Reliability and statistics in geotechnical engineering. Wiley and Sons, 605 p

3.   Bahat D, Grossenbacher K, Karasaki K (1999) Mechanism of exfoliation joint formation in granitic rocks, Yosemite National Park. J Struct Geol 21:85–96

4.   Bandis SC (1980) Experimental studies of scale effects on the shear strength, and deformation of rock joints. Unpublished PhD thesis, The University of Leeds, 385 p plus 4 appendices

5.   Bandis SC, Lumsden AC, Barton NR (1981) Experimental studies of scale effects on the shear behaviour of rock joints. Int J Rock Mech Min Sci Geomech Abstr 18:1–21

6.   Barton NR (1973) Review of a new shear-strength criterion for rock joints. Eng Geol 7:287–332

7.   Barton NR (1990) Scale effects or sampling bias? In: Proceedings of 1st international workshop on scale effects in rock masses. Loen, Norway, pp 31–55

8.   Barton NR, Bandis SC (1990) Review of predictive capabilities of JRC-JCS model in engineering practice. In: Barton N, Stephansson O (eds) Rock joints, proceedings international symposium on rock joints, Loen, Norway. Balkema, Rotterdam, pp 603–610

9.   Beer AJ, Stead D, Coggan JS (2002) Estimation of the joint roughness coefficient (JRC) by visual comparison. Rock Mech Rock Eng 35:65–74

10. Brady BHG, Brown ET (1985) Rock mechanics for underground mining.

George Allen and Unwin, London, 527 p

11. Brand EW, Hencher SR, Youdan JD (1983) Rock slope engineering in Hong Kong. In: Proceedings of the 5th international rock mechanics congress, Melbourne, C, 17–24

12. Brown RHA (1999) The management of risk in the design and construction of tunnels. Korean Geotechnical Society, Tunnel Committee Seminar, 21st Seminar, 21st September, 22 p

13. Brown ET (2008) Estimating the mechanical properties of rock masses. In: Potvin Y, Carter J, Dyskin A, Jeffrey R (eds) Proceedings, 1st southern hemisphere international rock mechanics symposium, Perth, 16–19 September, vol 1, pp 3–22

14. BS 5930 (1999) Code of practice for site investigation. 206 p

15. Byerlee J (1978) Friction of rocks. Pure Appl Geophys 116:615–626

16. Carter TG, Mierzejewski J, Kwong AKL (1998) Site investigation for rock slope excavation and stabilization adjacent to a major highway in Hong Kong. In: Proceedings international conference on urban ground engineering, session 2, Geotechnics, Paper 14, 10 pp

17. Carter TG, de Graaf P, Booth P, Barrett S, Pine R (2002) Integration of detailed field investigations and innovative design key factors to the successful widening of the Tuen Mun Highway. In: Proceedings of the 22nd annual seminar, organised by the Geotechnical Division of the Hong Kong Institution of Engineers, 8th May, 2002, pp 187–201

18. Carter TG, Diederichs MS, Carvalho JL (2008) Application of modified Hoek–Brown transition relationships for assessing strength and post-yield behaviour at both ends of the rock competence scale. Proc SAIMM 108(6):325–338

19. Carvalho JL, Carter TG, Diederichs MS (2007) An approach for prediction of strength and post yield behaviour for rock masses of low intact strength. In: Proc 1st Can–US Rock Symp Meeting society's challenges and demands. Vancouver, pp 249–257

20. Cloos H (1922) Tektonik und magma. Bd I Abh d Preuss Geol Land, 89

21. Diederichs MS (2003) Rock fracture and collapse under low confinement conditions. Rock Mech Rock Eng 36:339–381

22. El-Ramly H, Morgenstern NR, Cruden DM (2005) Probabilistic assessment of stability of a cut slope in residual soil. Géotechnique 55(1):77–84

23. Fecker E, Rengers N (1971) Measurement of large scale roughness of rock planes by means of profilograph and geological compass. In:

Proceedings symposium on rock fracture, Nancy, France, Paper 1–18

24. Geotechnical Engineering Office (1988) Guide to rock and soil descriptions. (Geoguide 3). Geotechnical Engineering Office, Hong Kong, 189 p. Available at http://www.cedd.gov.hk/eng/publications/manuals/manu_eg3.htm

25. Geotechnical Engineering Office (2000a) Highway slope manual. Geotechnical Engineering Office, Hong Kong, 114 p. Available at http://www.cedd.gov.hk/eng/publications/manuals/manu_em2.htm

26. Geotechnical Engineering Office (2000b) Technical guidelines on landscape treatment and bio-engineering of man-made slopes and retaining walls. GEO Publication No. 1/2000, Geotechnical Engineering Office, Hong Kong, 146 p. Available at http://www.cedd.gov.hk/eng/publications/manuals/manu_ep1.htm

27. Goodman RE (1980) Introduction to rock mechanics. Wiley and Sons, New York

28. Hack R (1998) Slope stability probability classification, SSPC, 2nd edn. International Institute for Aerospace Survey and Earth Sciences (ITC), Publication No. 43, 258 p

29. Halcrow Asia Partnership Ltd (1998) Report on the landslide at Ten Thousand Buddha's Monastery of 2 July 1997. GEO Report No. 77, Geotechnical Engineering Office, Hong Kong, 96 p. http://www.cedd.gov.hk/eng/publications/geo_reports/geo_rpt077.htm

30. Halcrow China Ltd (2002a) Investigation of some selected landslides in 2000 (Volume 1). GEO Report No. 129, Geotechnical Engineering Office, Hong Kong, 144 p. Available at http://www.cedd.gov.hk/eng/publications/geo_reports/geo_rpt129.htm

31. Halcrow China Ltd (2002b) Investigation of some selected landslides in 2000 (Volume 2). GEO Report No. 130, Geotechnical Engineering Office, Hong Kong, 173 p. Available at http://www.cedd.gov.hk/eng/publications/geo_reports/geo_rpt130.htm

32. Halcrow China Ltd (2002c) Study on the provision of temporary protective barriers and associated measures and methods of and the supervision to be specified and provided for rock breaking operations. Unpublished Report to the Geotechnical Engineering Office, Hong Kong Government

33. Hencher SR (1976) Correspondence: a simple sliding apparatus for the measurement of rock friction. Géotechnique 26(4):641–644

34. Hencher SR (1983) Landslide studies 1982 case study no. 2 Junk Bay

Road. Special project report no SPR 3/83, Geotechnical Control Office, Hong Kong, 32 p

35. Hencher SR (1995) Interpretation of direct shear tests on rock joints. In: Daeman JJK, Schultz RA (eds) Proceedings 35th US symposium on rock mechanics. Lake Tahoe, pp 99–106

36. Hencher SR (2006) Weathering and erosion processes in rock—implications for geotechnical engineering. In: Proceedings symposium on Hong Kong soils and rocks, March 2004, Institution of Mining, Metallurgy and Materials and Geological Society of London, pp 29–79

37. Hencher SR (2010) Preferential flow paths through soil and rock and their association with landslides. Hydrol Process (in press)

38. Hencher SR, Knipe RJ (2007) Development of rock joints with time and consequences for engineering. In: Proceedings of the 11th congress of the international society for rock mechanics, vol 1, pp 223–226

39. Hencher SR, Richards LR (1982) The basic frictional resistance of sheeting joints in Hong Kong Granite. Hong Kong Eng 11(2):21–25

40. Hencher SR, Richards LR (1989) Laboratory direct shear testing of rock discontinuities. Ground Eng 22(2):24–31

41. Hencher SR, Toy JP, Lumsden AC (1993) Scale dependent shear strength of rock joints. In: da Cunha P (ed) Scale effects in rock masses 93. Balkema, Rotterdam, pp 233–240

42. Hoek E (1968) Brittle failure of rock. In: Stagg KG, Zienkiewicz OC (eds) Rock mechanics in engineering practice. Wiley and Sons, London, pp 99–124

43. Hoek E (2009) A slope stability problem in Hong Kong. Published online at http://www.rocscience.com/hoek/pdf/7_A_slope_stability_problem_in_Hong_Kong.pdf

44. Holzhausen GR (1989) Origin of sheet structure, 1. Morphology and boundary conditions. Eng Geol 27:225–278

45. International Society for Rock Mechanics (1978) Suggested methods for the quantitative description of discontinuities in rock masses. Int J Rock Mech Min Sci Geomech Abstr 15(6):319–368

46. Jahns RH (1943) Sheet structure in granites, its origin and use as a measure of glacial erosion in New England. J Geol 51:71–98

47. Johnson AM (1970) Formation of sheet structure in granite, chap 10. In: Physical processes on geology, Freeman Cooper and Company

48. Kikuchi K, Mito Y (1993) Characteristics of seepage flow through the actual rock joints. In: Proceedings 2nd international workshop on scale

effects in rock masses, Lisbon, pp 305–312

49. Kulatilake PHSW, Shou G, Huang TH, Morgan RM (1995) New peak shear strength criteria for anisotropic rock joints. Int J Rock Mech Min Sci Geomech Abstr 32:673–697

50. Kveldsvik V, Nilsen B, Einstein HH, Nadim F (2008) Alternative approaches for analyses of a 100,000 m³ rock slide based on Barton–Bandis shear strength criterion. Landslides 5:161–176

51. Martel SJ (2006) Effect of topographic curvature on near-surface stresses and application to sheeting joints. Geophys Res Lett 33:LO1308

52. Nichols T C Jr (1980) Rebound, its nature and effect on engineering works. Q J Eng Geol 13:133–152

53. Nicholson GA (1994) A test is worth a thousand guesses—a paradox. In: Nelson PP, Laubach SE (eds) Proc 1st NARMS symposium, pp 523–529

54. Ollier CD (1975) Weathering. Longman, 304 p

55. Papaliangas TT, Hencher SR, Lumsden AC (1994) Scale independent shear strength of rock joints. In: Proceedings IV CSMR/integral approach to applied rock mechanics, Santiago, Chile, pp 123–133

56. Papaliangas TT, Hencher SR, Lumsden AC (1995) A comprehensive peak strength criterion for rock joints. In: Proceedings 8th international congress on rock mechanics, Tokyo, vol 1, pp 359–366

57. Parry S, Campbell SDG, Fletcher CJN (2000) Kaolin in Hong Kong saprolites—genesis and distribution. In: Proceedings conference engineering geology HK 2000, IMM HK Branch, pp 63–70

58. Patton FD (1966) Multiple modes of shear failure through rock. In: Proceedings 1st international congress international society rock mechanics, Lisbon, vol 1, pp 509–513

59. Patton FD, Deere DU (1970) Significant geological factors in rock slope stability. In: Proceedings symposium on planning open pit mines, Johannesburg, A.A. Balkema, pp 143–151

60. Pine RJ, Roberds WJ (2005) A risk-based approach for the design of rock slopes subject to multiple failure modes—illustrated by a case study in Hong Kong. Int J Rock Mech Min Sci 42:261–275

61. Rabinowicz E (1965) Friction and wear of materials. John Wiley, New York

62. Richards LR, Cowland JW (1982) The effect of surface roughness on the field shear strength of sheeting joints in Hong Kong granite. Hong Kong Eng 10(10):39–43

63. Richards LR, Cowland JW (1986) Stability evaluation of some urban rock

slopes in a transient groundwater regime. In: Proceedings conference on rock engineering and excavation in an urban environment, IMM, Hong Kong, pp 357–363 (discussion 501–506)

64. Roorda J, Thompson JC, White OL (1982) The analysis and prediction of lateral instability in highly stressed, near-surface rock strata. Can Geotech J 19(4):451–462

65. Schneider HJ (1976) The friction and deformation behaviour of rock joints. Rock Mech 8:169–184

66. Scholtz CH (1990) The mechanics of earthquakes and faulting. Cambridge University Press, Cambridge, 439 p

67. Selby MJ (1993) Hillslope materials and processes, 2nd edn, Oxford University Press, 451 p

68. Simons N, Menzies B, Mathews M (2001) A short course in soil and rock slope engineering. Thomas Telford, 432 p

69. Spang K, Egger P (1990) Action of fully-grouted bolts in jointed rock and factors of influence. Rock Mech Rock Eng 23:201–229

70. Stimpson B (1981) A suggested technique for determining the basic friction angle of rock surfaces using core. Int J Rock Mech Min Sci Geomech Abstr 18:63–65

71. Twidale CR (1973) On the origin of sheet jointing. Rock Mech 5:163–187

72. Twidale CR, Vidal Romani JR (2005) Landforms and geology of granite terrains. Taylor and Francis, 352 p

73. Vidal Romani JR, Twidale CR (1999) Sheet fractures, other stress forms and some engineering implications. Geomorphology 31:13–27

74. Wakasa S, Matsuzaki H, Tanaka Y, Matskura Y (2006) Estimation of episodic exfoliation rates of rock sheets on a granite dome in Korea from cosmogenic nuclide analysis. Earth Surf Proc Land 31:1246–1256

75. Wise DU (1964) Microjointing in basement, Middle Rocky Mountains of Montana and Wyoming. Geol Soc Am Bull 75:287–306

76. Wyllie DC, Mah CW (2004) Rock Slope Engineering, 4th edn, Spon, 431 p

77. Yu YS, Coates DF (1970) Analysis of rock slopes using the finite element method. Department of Energy, Mines and Resources Mines Branch, Mining Research Centre, Research Report, R229 (Ottawa)

78. Yu YF, Siu CK, Pun WK (2005) Guidelines on the use of prescriptive measures for rock slopes. GEO Report 161, 34 p

Chapter 6

EVALUATION OF MODE I FRACTURE TOUGHNESS ASSISTED BY THE NUMERICAL DETERMINATION OF K-RESISTANCE

Takahiro Funatsu[1], Norikazu Shimizu[2], Mahinda Kuruppu[3], Kikuo Matsui[4]

[1]National Institute of Advanced Industrial Science and Technology (AIST), Central 7, 1-1-1 Higashi, Tsukuba, Ibaraki 305-8567, Japan

[2]Yamaguchi University, 2-16-1 Tokiwadai, Ube, Yamaguchi 755-8611, Japan

[3]Curtin University, Locked Bag 30, Kalgoorlie, WA 6433, Australia

[4]Kyushu University, 744 Motooka, Nishi-ku, Fukuoka 819-0395, Japan

ABSTRACT

The fracture toughness of a rock often varies depending on the specimen shape and the loading type used to measure it. To investigate the mode I fracture toughness using semi-circular bend (SCB) specimens, we experimentally studied the fracture toughness using SCB and chevron bend (CB) specimens, the latter being one of the specimens used extensively as an International Society for Rock Mechanics (ISRM) suggested method, for comparison. The mode I fracture toughness measured using SCB specimens is lower than both the level I and level II fracture toughness values measured using CB specimens. A numerical study based on discontinuum mechanics was conducted using a two-dimensional distinct element method (DEM) for evaluating crack propagation in the SCB specimen during loading. The numerical results indicate subcritical crack growth as well as sudden crack propagation when the load reaches the maximum. A K-resistance curve is drawn using the crack extension and the load at the point of evaluation. The fracture toughness evaluated by the K-resistance curve is in agreement with the level II fracture toughness measured using CB specimens. Therefore, the SCB specimen yields an improved value for fracture toughness when the increase of K-resistance with stable crack propagation is considered.

INTRODUCTION

In rock engineering problems dealing with the stability of structures, controlling crack initiation and propagation is very important. Microcracks and macrocracks affect the rock mass strength and deformation; these factors strongly influence the stability of geological structures such as underground and open pit mines, tunnels, and rock slopes. Rock fracturing also plays a key role in the exploitation of energy resources in that creating new cracks enhances the production of oil, natural gas, and geothermal energy, and also facilitates leakage paths in the sequestration of CO_2 in geological storage sites.

The fracture toughness is a measure of a material's resistance to crack propagation. The fracture toughness of rock materials has been determined using various test specimen conFigureurations and methods. The International Society for Rock Mechanics (ISRM) has incorporated chevron bend (CB) (ISRM 1988), short rod (SR) (ISRM 1988), and cracked chevron-notched Brazilian disk (CCNBD) (Fowell 1995) specimens into the standard method for the measurement of the fracture toughness of rock materials. Three-point bending-type specimens such as single edge cracked round bar bend (SECRBB) (Ouchterlony 1981), semi-circular bend (SCB) (Chong and Kuruppu 1988), and straight notched disk bending (SNDB) (Tutluoglu and Keles2011) specimens, as well as Brazilian disk-type specimens such as cracked straight-through Brazilian disk (CSTBD) (Fowell and Xu 1994) and flattened Brazilian disk (Wang and Xing 1999) specimens, have also been used for the measurement of the fracture toughness. Among these, the fracture toughness measurement method using the SCB specimen shown in Figure. 1 has been recently approved as an ISRM suggested method (Kuruppu et al. 2014). It is a core-based specimen that possesses inherently favorable characteristics such as simplicity, minimal machining requirements, and easy testability through the application of three-point compressive loading using a standard test frame.

It is observed that, for the same rock sample, the mode I fracture toughness varies when different specimen types are used for measurement. Chang et al. (2002) measured the fracture toughness of granite and marble using CB, CCNBD, SCB, chevron-notched SCB, and uncracked Brazilian disk test (BDT) (Guo et al. 1993) specimens. The mode I fracture toughness values of granite and marble measured using SCB specimens were 0.68 ± 0.19 MPam$^{0.5}$ with 31 specimens and 0.87 ± 0.15 MPam$^{0.5}$ with 27 specimens, respectively.

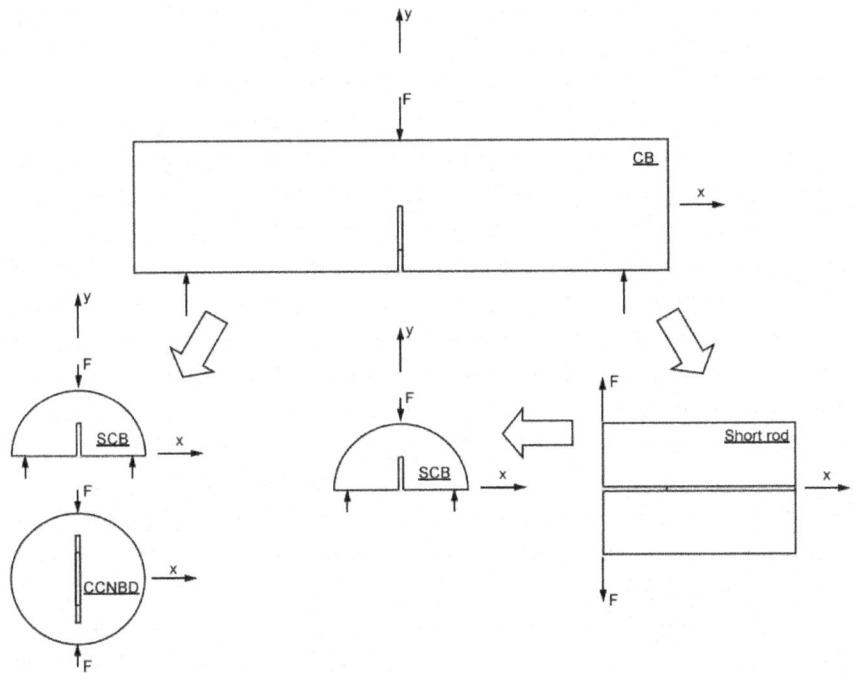

Figure. 1: Core-based fracture toughness test specimens illustrating their application to anisotropic materials [modified from Chong and Kuruppu (1988)]

These values are lower than the fracture toughness values measured using other specimens. Khan and Al-Shayea (2000) measured the fracture toughness of limestone using SCB, CSTBD, CCNBD, and SECRBB specimens; the average values measured using the SCB, CSTBD, CCNBD, and SECRBB specimens were 0.68, 0.42, 0.61, and 0.55 MPam$^{0.5}$, respectively. Tutluoglu and Keles (2011) found that the fracture toughness values of andesite measured using CCNBD, SCB, and SNDB specimens were 1.45 ± 0.06 MPam$^{0.5}$ with five specimens, 0.94 ± 0.12 MPam$^{0.5}$ with 21 specimens, and 1.00 ± 0.09 MPam$^{0.5}$ with 20 specimens, respectively. They argued that the variation of the fracture toughness was due to the differences in the size of the fracture process zone (FPZ). They evaluated the size of the FPZ of SCB and SNDB specimens and found that the size of the FPZ of the former was 2.15 times larger than that of the FPZ of the latter. Aliha et al. (2012) reported that the fracture toughness strongly depends on the geometry and loading conditions of the test specimen. They showed that the fracture toughness of Guiting limestone measured using an SCB specimen was higher than that measured using a CSTBD specimen, and they discussed the difference between the values using the maximum tangential stress (MTS) criterion. They found that the higher-order stress term A_3 was responsible for the variation of the fracture toughness. Iqbal and Mohanty (2007) showed that the fracture toughness values of CB and CCNBD specimens for a brittle rock were comparable.

These studies indicate that the fracture toughness must be dealt with carefully if it is considered representative of the rock material, especially if a non-ISRM suggested method is adopted. The size of the FPZ and pre-critical stable crack growth are the key factors affecting the fracture toughness. Nasseri et al. (2006) measured the fracture toughness and acoustic emission activity in brittle rocks. They found that the variation of fracture toughness is caused by the pre-existing microcrack density and its orientation with respect to the fracture propagation direction. The creation of an FPZ surrounding the propagating main crack has been confirmed by acoustic emission techniques. Dai et al. (2007) investigated the effect of crack–microcrack interaction on the anisotropic behavior of fracture toughness. The microstructural investigation of thin sections indicated that the pre-existing microcracks caused the variation of the fracture toughness values. The FPZ or crack growth can be estimated in several ways. Optical methods are used to observe moiré fringe patterns during loading and to measure the size of the FPZ. Acoustic emission measurement is also used to estimate the size of the FPZ. The compliance is used to indirectly measure the crack growth during loading. Apart from laboratory studies, numerical modeling is also useful for estimating the crack growth. We investigated the application of the distinct element method (DEM) (Cundall and Strack 1979), which is based on discontinuum mechanics, because crack propagation and microcracking occur in a discontinuous manner. DEM has been used to study crack propagation in rocks or rock-like materials such as concrete. For example, Azevedo and Lemos (2006) presented a DEM/finite element method (FEM) coupling algorithm which enables DEM to be used in the discretization of the fracture zone and for the surrounding areas of a discretization based on the FEM. The hybrid DEM/FEM method was applied for fracture analysis in concrete. They successfully modeled the crack localization process and pre-peak load versus displacement curves of both mode I and mixed mode fracture experiments performed using beam specimens. Tan et al. (2009) used the DEM software package PFC2D (Itasca Consulting Group Inc. 2004) for modeling the fracture and damage processes of polycrystalline silicon carbide (SiC) ceramics. They modeled the fracture toughness testing using a specimen subjected to three-point bend loading and showed that the numerical results agreed with the experimental measurements. D'Addetta et al. (2002) presented a combined particle and lattice model as an improved DEM formulation. It was applied to model the fracture process of cohesive granular materials somewhat similar to sandstone. They were successful in showing the typical microcrack nucleation, growth, and coalescence to form macrocracks under tension, compression, and shear modes of loading. Their simulation results were in agreement with the experimental observations. These studies demonstrate that DEM can be used for simulating the mechanical behavior of rocks and crack propagation behavior, and that it is a useful tool for investigating the FPZ and crack growth during loading.

In this study, we investigated the mode I fracture toughness using an SCB specimen. Furthermore, we used a CB specimen for the purpose of comparison of the fracture toughness. A DEM model of the SCB test specimen was used to investigate the crack growth and FPZ during loading. We evaluated the mode I fracture toughness using the SCB specimen and its K-resistance curve and found that the corrected mode I fracture toughness is comparable to the level II fracture toughness measured using the CB specimen.

METHODOLOGY

Test Material

Kimachi sandstone produced in Shimane Prefecture, Japan, was used as the test material. The mechanical properties of the rock are listed in Table 1. An analysis of this material using X-ray diffractometry (XRD) revealed that it mainly consists of albite, anorthite, quartz, montmorillonite, and mordenite.

Table 1: Mechanical properties of Kimachi sandstone

σ_c (MPa)	E_{50} (GPa)	v	σ_t (MPa)
66.9	13.2	0.18	4.9

Kimachi sandstone has been found to be slightly anisotropic (Funatsu et al. 2004). The principal directions of anisotropy are known as arrester, divider, and short-transverse. All tests using SCB and CB specimens reported herein were performed in the arrester orientation.

Experimental Method

Testing

The tests were carried out using the SCB specimen configuration shown in Figure. 1. Such an SCB specimen can be made from leftover core material after testing CB or SR specimens, so that the variation of the material properties of the rock is kept to a minimum. This specimen has certain inherently favorable properties such as simplicity, minimal machining requirements, and easy testability by means of three-point compressive loading using a standard test frame (Figure. 2).

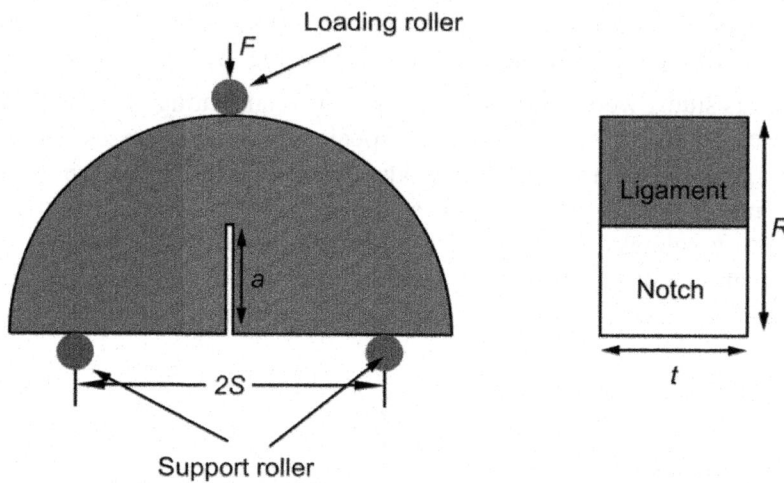

Figure. 2: Semi-circular bend (SCB) specimen geometry and schematic loading arrangement (R radius of the specimen, t thickness, a notch length, 2S distance between the two supporting pins, F monotonically increasing compressive load applied at the central loading pin of the three-point bend loading).

Specimens were prepared by sawing or slicing rock cores that were drilled in the direction of bedding planes. Each resulting disk was then cut into two halves, along a plane parallel to the direction of the bedding planes, to form two specimens. Specimens of 50-mm radius and 25-mm thickness were used. A straight notch was introduced in each specimen using a diamond circular saw, such that the notch-length-to-radius ratio was 0.3, 0.4, or 0.5. The thickness of the saw blade used was 0.3 mm, yielding a notch of similar thickness. The resulting SCB test specimens had their notches in the arrester orientation with respect to the material anisotropy (Chong et al. 1987). The specimens were oven-dried at 40 °C for 120 h, and all dimensions were recorded.

The specimens were placed on the loading platform such that the span ratio S/R was 0.8 and then tested to failure under load-line displacement control and at a loading rate of 0.075 mm/min (see Figure. 3). The load, load-point displacement (LPD), and crack-opening displacement (COD) were recorded as functions of time during each test.

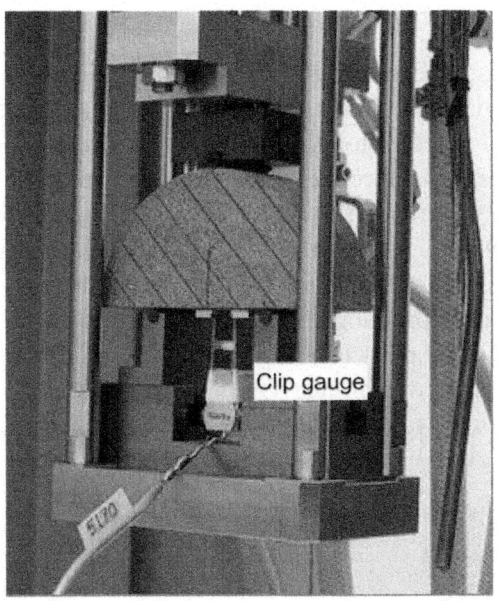

Figure. 3: Sample setup in the loading frame with a crack-opening displacement (COD) gauge.

Derivation of Fracture Toughness by the SCB Specimen

The fracture toughness, K_{Ic}, is determined using the peak load, non-dimensional stress intensity factor, and specimen dimensions. For the SCB specimens, K_{SCB} is given as (Chong et al. 1987):

$$K_{SCB} = Y\sigma_0\sqrt{\pi a} \tag{1}$$

where Y is the normalized stress intensity factor, $\sigma_0 = F_{max}/2Rt$, F_{max} is the maximum load, a is the notch length, R is the specimen radius, and t is the thickness. The stress intensity factor Y is a function of the a/Rratio, β, and the half-span-to-radius ratio S/R. The best-fit curve for Y is given by Lim et al. (1994) as:

$$Y = \frac{S}{R}(2.91 + 54.39\beta - 391.4\beta^2 + 1210.6\beta^3$$
$$-1650\beta^4 + 875.9\beta^5) \tag{2}$$

where S is the half-span length of the support rollers. Equation (2) is valid for $0.1 \le \beta \le 0.8$. An S/R ratio of 0.8 was chosen for the fracture toughness tests performed with the SCB specimen.

CB Specimen

Fracture toughness measurement by a CB specimen is one of the ISRM suggested methods. The evaluation of fracture toughness is done at two levels. The level I fracture toughness based on fracture load is suitable if the testing material is considered as a linear elastic material. The curve of the normalized stress intensity factor versus notch length for a chevron notch has a minimum value, suggesting that the initial crack growth occurs stably and that the specimen fails upon reaching the minimum value of the stress intensity factor corresponding to the maximum applied load. Therefore, the minimum normalized stress intensity factor is used for evaluating the level I fracture toughness. For nonlinearly behaving materials, the level II fracture toughness corrects the level I fracture toughness by considering the degree of nonlinearity p. Figure 4 shows a typical load versus LPD curve.

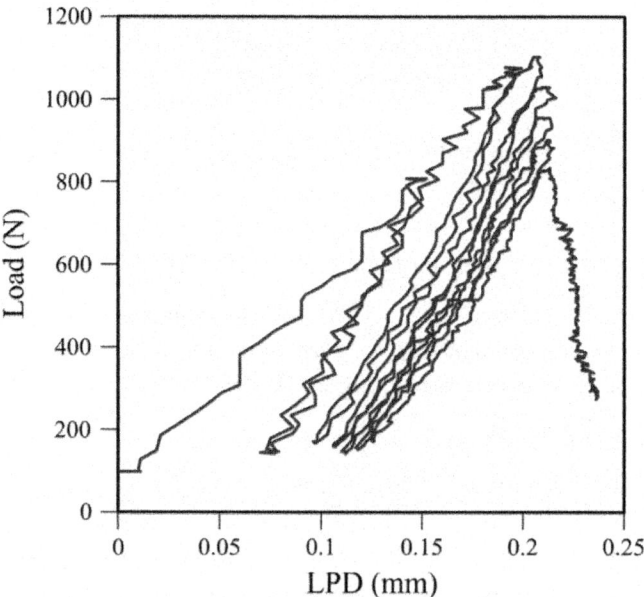

Figure. 4: Load versus load-point displacement (LPD) curve obtained by the chevron bend (CB) test also showing partial unloading cycles to facilitate determining the level II fracture toughness.

The level I fracture toughness, K_{CB}, can be calculated by the following equation (ISRM 1988):

$$K_{CB} = A_{min}F_{max}/D^{1.5} \qquad (3)$$

where F_{max} is the maximum load, D is the diameter of the specimen, and:

$$A_{\min} = \left[1.835 + 7.15a_0/D + 9.85(a_0/D)^2\right]2S/D \tag{4}$$

where S is the half-span length between support points and a_0 is the chevron tip distance from the specimen surface.

The level II fracture toughness can be calculated as:

$$K_{CB}^c = \sqrt{(1+p)/(1-p)}F_c/F_{\max}K_{CB} \tag{5}$$

where p is the degree of nonlinearity and F_c is the load at the evaluation point. Here, p and F_c are determined from the load versus COD curve (ISRM 1988). We used D = 60 mm, a_0 = 9 mm, andS = 99.9 mm in accordance with the suggested method (ISRM 1988) and the specimens were made with their notches in the arrester orientation with respect to the bedding planes.

Numerical Model by DEM

Analytical methods based on continuum mechanics are normally used in the design of many geological structures, such as roadway tunnels. However, it is difficult to use continuum mechanics to simulate failures like the separation of materials and shear planes. In this study, a DEM-based two-dimensional discontinuum program called PFC2D (Itasca Consulting Group Inc. 2004) is used to simulate crack propagation in rocks. PFC2D simulates the mechanical behavior of a material by representing it as an assemblage of circular particles that can be bonded to one another. The basic mechanical properties such as Young's modulus and Poisson's ratio are derived from laboratory tests. In the continuum model, the elastic properties can be used directly. However, in PFC2D, the mechanical behavior of the assemblage is dominated by the microproperties of the particles and the bonds between them. These microproperties cannot be determined from laboratory tests. Thus, the relationship between the microproperties and the macroproperties should be determined by the modeling of rock testing, such as the uniaxial compressive strength (UCS) test and the Brazilian test, prior to the simulation of fracture toughness. Moreover, a clumped particle model (Cho et al.2007), which combines particles located within a circle, is adopted in this study; each clump behaves as an element having a complicated shape.

Numerical Modeling of the Uniaxial Compressive and Brazilian Tests

Simulations of the uniaxial compressive and Brazilian tests were conducted to calibrate the appropriate input parameters. These simulations were performed according to the works of Potyondy and Cundall (2004). The specimen for the compressive tests is 120 mm in length and 60 mm in width, and the diameter of the specimen for the Brazilian tests is 60 mm. Both the particles and models themselves have thicknesses of one unit of length, which is equal to 1 m. The calibration process is explained in detail by Funatsu et al. (2008). The number of particle elements is about 21,000 for the uniaxial compressive test model. Being a two-dimensional code, the PFC2D is unable to simulate the compressive test of a cylindrical specimen. Therefore, we decided to simulate the compressive test of a rectangular specimen having unit thickness.

For the uniaxial compressive tests, the top and bottom wall elements act as the loading platens, and the velocity of the walls is kept constant at 5.0×10^{-5} mm/step (i.e., the rate of load application). For the Brazilian tests, the side walls act as platens, and the velocity of the walls is kept constant at 5.0×10^{-5} mm/step. The models for the uniaxial compressive tests and Brazilian tests are shown in Figure. 5. The applied stress is measured by dividing the average force acting on opposing walls by the area of the corresponding specimen cross-section (Potyondy and Cundall 2004).

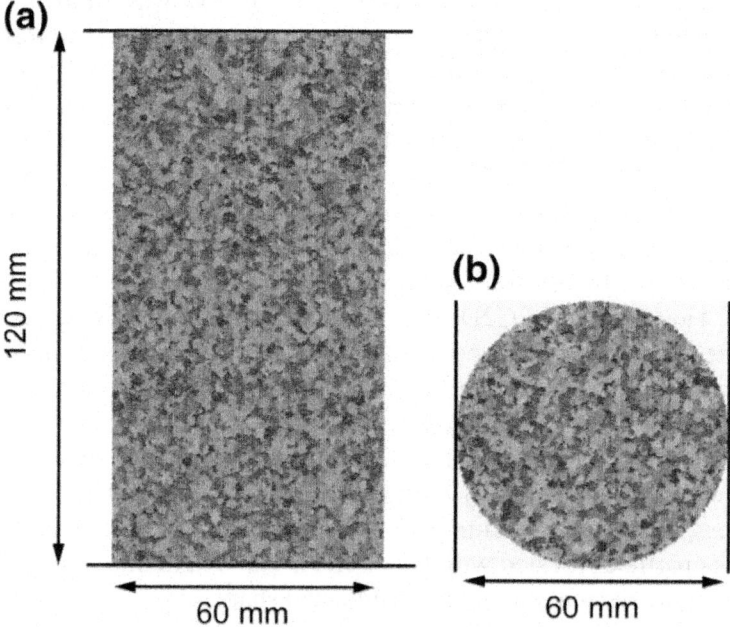

Figure. 5: Distinct element method (DEM) models of: **a** the uniaxial compressive test and **b** the Brazilian test (Funatsu et al. 2008).

Figures 6 and 7 show the stress–strain curves derived by the physical tests, along with the results of the simulating UCS and Brazilian tests, respectively. With the appropriate selection of DEM parameters, it can be seen that the model simulates the experimental results very well. This is despite the fact that a two-dimensional rectangular model was used, which is expected to be weaker in compression than if a cylindrical model was used. The crack distribution indicates that tensile failure occurs along the loading direction. The tensile strength calculated by the numerical model of the Brazilian test is 5.0 MPa. The difference in tensile strength between the numerical simulation and the experiments is only 0.1 MPa. However, there is a limitation of the numerical model in comparison with the test specimen. The former is a two-dimensional model and the latter is a three-dimensional cylindrical shape. Cho et al. (2007) showed that the two-dimensional clumped particle model can reproduce the failure envelope of both hard rock and weak rock. Our results support their findings. A set of input parameters suitable for the modeling of Kimachi sandstone is given in Table 2.

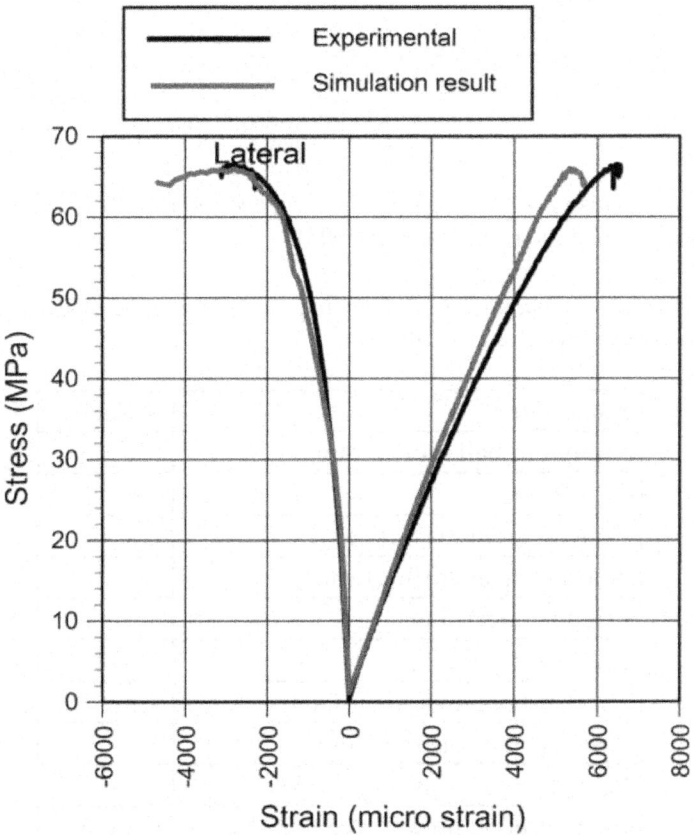

Figure. 6: Comparison of the stress–strain curves of the uniaxial compressive test; experimental and numerical simulation (Funatsu et al. 2008).

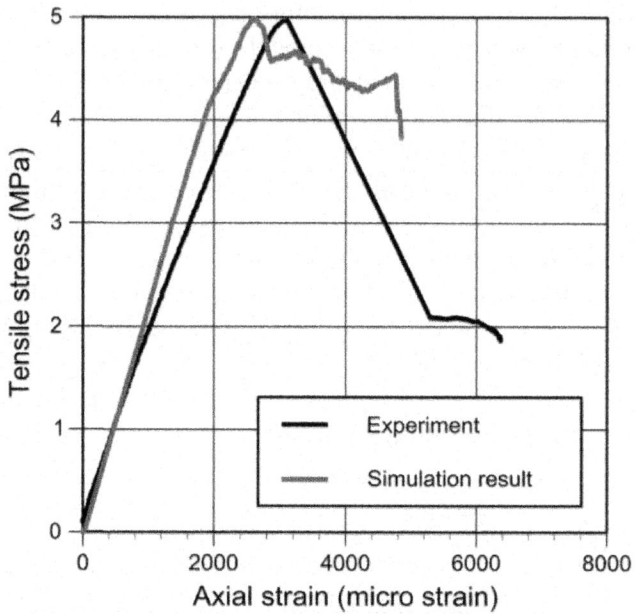

Figure. 7: Comparison of the stress–strain curves of the Brazilian test; experimental and numerical simulation (Funatsu et al. 2008).

Table 2: Microscopic parameters for distinct element method (DEM) modeling of Kimachi sandstone (Funatsu et al.2008)

Parameter	Value
Minimum ball radius (mm)	0.2
Ball size ratio	1.5
Contact modulus (GPa)	2.8
Normal/shear stiffness ratio	1.5
Friction coefficient	0.2
Ball density (kg/m³)	2,630
Parallel-bond modulus (GPa)	2.8
Parallel-bond stiffness ratio	1.5
Parallel-bond radius multiplier	1
Parallel-bond normal strength (MPa)	6 ± 0.6
Parallel-bond shear strength (MPa)	55 ± 5.5
Clump radius (mm)	1.0 ± 0.2

Modeling of the SCB Specimen by DEM

The model of the SCB specimen is shown in Figure. 8b. Figure 8a shows the SCB specimen for comparison. The notch is created by deleting the particles located within the notch. The specimen diameter is 100 mm and initial notch length is 25 mm. The thickness of the test specimen is 25 mm. The numerical model is of unit length thickness (1 m). The support and loading rollers for three-point bending are created by wall elements. The support rollers are kept separated by a fixed span length of 80 mm, which is same as that used in the physical tests. The element size was defined as shown in Table 2. The particles were randomly packed with a uniform size distribution. The number of particles was 11,663. The loading roller located above the specimen is made to move downward at a constant displacement rate of 5.0×10^{-6} mm/step. The loading force used to evaluate the fracture toughness is taken as the force acting between the loading roller and the adjacent material particles. The displacement of the loading roller and the COD is also monitored. In PFC2D, stress cannot be calculated directly; instead, it is calculated as an average value inside the representative area, namely, the measurement circle. The radius of the circle is 3.0 mm, which is three times larger than the average radius of a clump. This diameter was selected to allow estimation of the proper stress state. The measurement circle was located in front of the initial notch tip. Since the stress calculated by the measurement circle is the average value inside the circle, the stress near the notch tip can be underestimated in the case of having a measurement circle with a large radius. In addition, if the crack extends during loading, the calculated stress change can be attenuated by the effect of averaging because the minimum crack extension is 0.2 mm, which is the same value as the minimum ball radius and is smaller than the radius of the measurement circle.

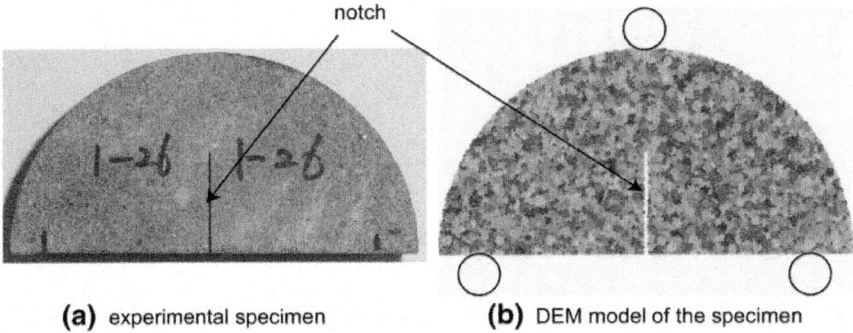

(a) experimental specimen   **(b)** DEM model of the specimen

Figure. 8: SCB specimen conFigureurations used for the **(a)** experimental and **(b)** DEM model. The specimen diameter Dis 100 mm, crack length a is 25 mm, thickness t is

25 mm for the test specimen and 1 m for the DEM model, and the span length between the two bottom supports 2S is 80 mm.

A microcracking in the numerical simulation is defined as a bond breakage between particles. The crack extension is defined as the length from the crack tip to the farther end of the connecting microcracks.

RESULTS

Fracture Toughness of Kimachi Sandstone Measured Experimentally Using SCB and CB Specimens

Figure 9 shows the test results for the fracture toughness using SCB specimens. The dots and bars in the Figure respectively indicate the average and standard deviation of the fracture toughness values for the same ratio of notch length to specimen radius. Three tests were performed at each crack length. This Figure shows that the fracture toughness measured using SCB specimens is independent of the notch length. The average value of fracture toughness using all data is 0.589 MPam$^{0.5}$, with a standard deviation of 0.0474 MPam$^{0.5}$.

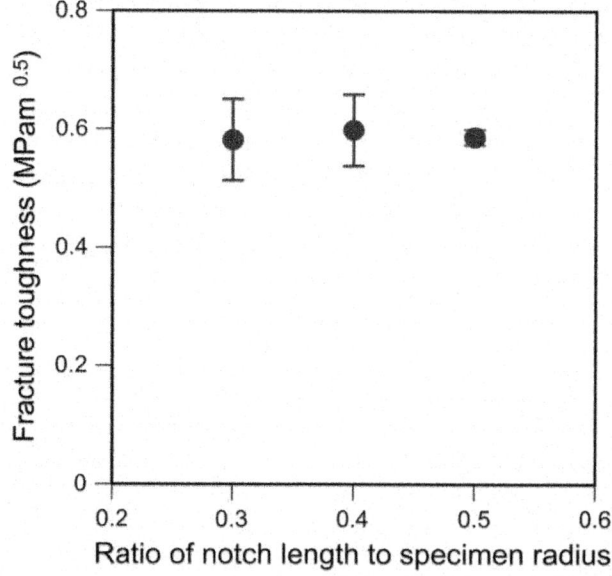

Figure. 9: Experimentally determined fracture toughness using SCB specimens.

The fracture toughness measured using CB specimens is summarized in Table 3. This table shows the level I fracture toughness K_{CB}, the level II fracture toughness $K_{CB}{}^{c}$, and the degree of nonlinearity p.

Table 3: Summary of fracture toughness test results using chevron bend (CB) specimens

Sample ID	K_{CB} (MPam$^{0.5}$)	p	$K_{CB}{}^c$ (MPam$^{0.5}$)
CB-1	0.781	0.176	0.954
CB-2	0.799	0.215	0.970
CB-3	0.798	0.271	1.053
CB-4	0.816	0.174	0.946
CB-5	0.781	0.149	0.885
Average	0.795	0.197	0.962
Standard deviation	0.0146	0.0476	0.060

The fracture toughness of Kimachi sandstone as measured using SR specimens was investigated by Matsuki et al. (1991). The level I fracture toughness K$_{SR}$ had values of 0.86 and 0.85 MPam$^{0.5}$, with specimen diameters of 80 and 100 mm, respectively. The level II fracture toughness K$_{SR}{}^c$ has values of 1.01 and 1.02 MPam$^{0.5}$, with specimen diameters of 80 and 100 mm, respectively. The fracture toughness measured using CB specimens in this study compares well with Matsuki et al.'s results. However, the level I and level II fracture toughness values measured using CB specimens are, respectively, ~35 and 63 % higher than those measured using SCB specimens.

Numerical Simulation of Fracture Toughness using the SCB Specimen

Figure 10 shows the load–COD curves determined by numerical modeling and through experimental results. The experimental result was from a typical test involving partial unloading cycles as shown in the Figure. Note that, in the simulation, the specimen thickness was set to 1 m. Therefore, the load shown in the Figure was converted from that corresponding to the thickness of 25 mm used in the experiment. The maximum load and the corresponding COD value were almost the same in the two sets of graphs. The fracture toughness determined by numerical modeling is 0.526 MPam$^{0.5}$, which is ~10 % lower than the average value of the fracture toughness measured experimentally using SCB specimens.

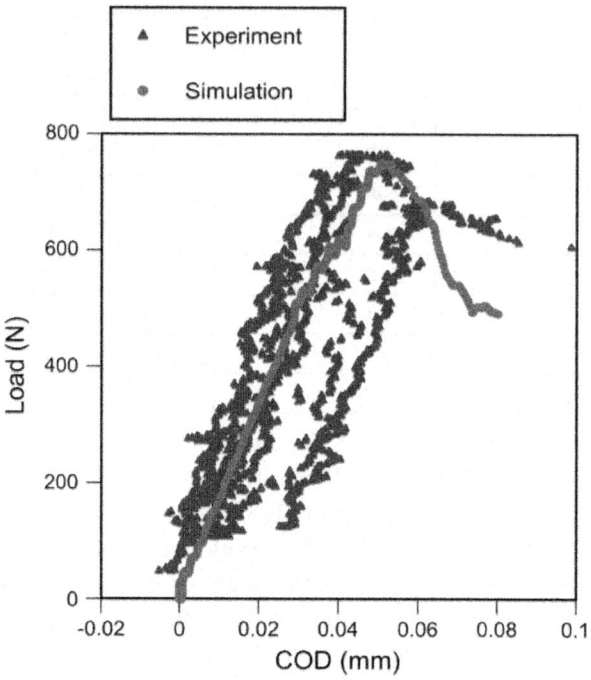

Figure. 10: Load versus COD curves obtained by numerical simulation and experimental methods. Note that the experimental graph includes partial unloading and reloading.

DISCUSSION

Mode I Fracture Toughness Measured by SCB and CB Tests

There are only a limited number of studies comparing the fracture toughness values measured using SCB specimens and ISRM suggested methods. Chang et al. (2002) compared the fracture toughness values determined using several different types of specimens, such as SCB, CB, CCNBD, and chevron-notched SCB specimens. The fracture toughness values of granite and marble measured using SCB specimens were 0.68 and 0.871 MPam$^{0.5}$, respectively. The level I fracture toughness values of granite and marble measured using CB specimens were ~1.4 and 1.1 MPam$^{0.5}$, respectively. These values are almost the same as those measured using CCNBD specimens. Tutluoglu and Keles (2011) reported that the level I fracture toughness values of andesite measured using SCB and CCNBD specimens were 0.94 and 1.45 MPam$^{0.5}$, respectively. Our results using sandstone showed that the trend is the same, i.e., the fracture toughness measured using SCB specimens is lower than that measured using CB and CCNBD specimens. One of the reasons for the difference in fracture toughness is the different notch type. Both the CB and CCNBD specimens have a chevron notch, whereas the SCB specimen has a straight notch. Furthermore, differences may exist in the development of the process zone, as it is affected by the type of loading, e.g., bending versus tension. Khan and Al-Shayea (2000) found that the fracture toughness values of limestone measured using CSTBD and CCNBD specimens were, respectively, 0.42 and 0.61 MPam$^{0.5}$. Chang et al. (2002) reported that the fracture toughness values measured using chevron-notched SCB specimens for granite and marble were 1.39 and 1.11 MPam$^{0.5}$, respectively. These values are ~2 times and 1.2 times higher than those measured using straight notched SCB specimens. In order to satisfy the requirements of LEFM on which the fracture toughness is based, the size of the FPZ during loading should be small enough so that the material behavior of the rock sample can be considered as linearly elastic. The radius of the FPZ can be derived from the following equation (Schmidt 1980):

$$r_c = \frac{1}{2\pi} \left(\frac{K_{Ic}}{\sigma_t} \right)^2 \qquad (6)$$

For Kimachi sandstone, the value of r_c calculated using the level II fracture toughness measured using CB specimens is 6.13 mm. The actual notch length and other dimensions need to be much larger than r_c. Furthermore, the fracture toughness may be underestimated unless (1) the crack starts to propagate when the FPZ is fully developed and (2) the fracture toughness is evaluated when the slow stable crack growth reaches its critical limit. The large difference in fracture toughness measured for CB and SCB specimens shows that the level I fracture toughness of the SCB specimen fails to satisfy those conditions.

Evaluation of Fracture Toughness by the *K*-Resistance Curve

When evaluating the fracture toughness, the crack growth that occurs at the critical level of loading cannot be ignored. If the crack growth is measured, the fracture toughness can be corrected as a function of the actual crack length using the K-resistance curve. In this study, we evaluated the crack growth by numerical modeling based on DEM. Figure 11 shows the load versus load-line displacement curve, as obtained by numerical modeling. The number of bond breakages corresponding to microcracking is also shown. The bond breakages suggest that a crack initiates before the load reaches the maximum value and that it propagates rapidly when the load reaches the maximum.

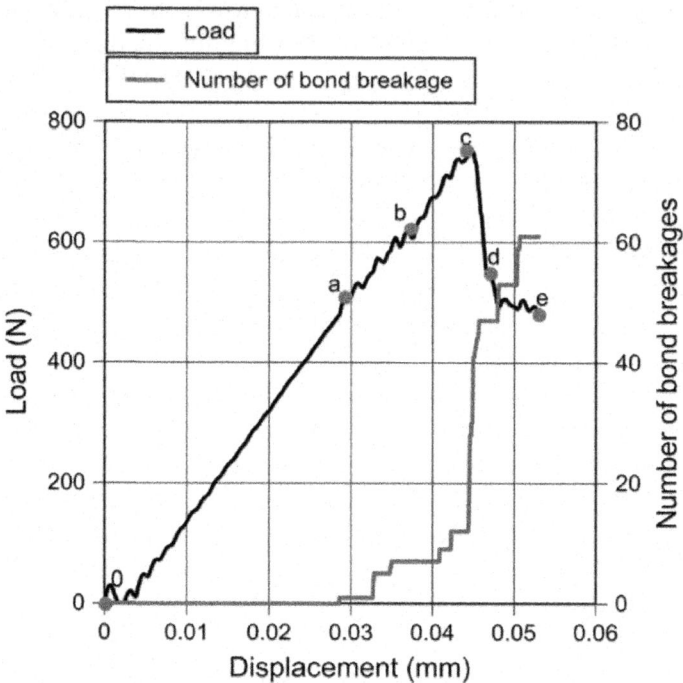

Figure. 11: Load versus displacement curve and cumulative number of bond break-ages for the case of a/R = 0.5 obtained by numerical simulation.

Figure 12 shows the crack growth during loading based on the numerical simulation. The crack growth at each point and the corresponding load can be derived from the numerical modeling. Parts a, b, c, d, and e in Figure. 12 correspond to points a, b, c, d, and e in Figure. 11. The crack growth at the maximum load is ~6 mm, as given in Figure. 12c. Furthermore, Figure. 12d shows the occurrence of sudden, unstable crack growth when the maximum load is reached. Following Eq. (1), the mode I stress intensity factor K_I at each point can be calculated as:

$$K_I = YF\sqrt{\pi(a + \Delta a)}\Big/2Rt \tag{7}$$

where F is the load at each evaluation point and Δa is the crack extension at that point. Y is the normalized stress intensity factor that corresponds to the notch length a + Δa. Similarly, we conducted a numerical simulation of the crack growth of specimens having a/R = 0.3 and 0.4, and constructed the K-resistance curves. Figure 13 shows the DEM models of an SCB specimen having a/R = 0.3 and a/R = 0.4, respectively.

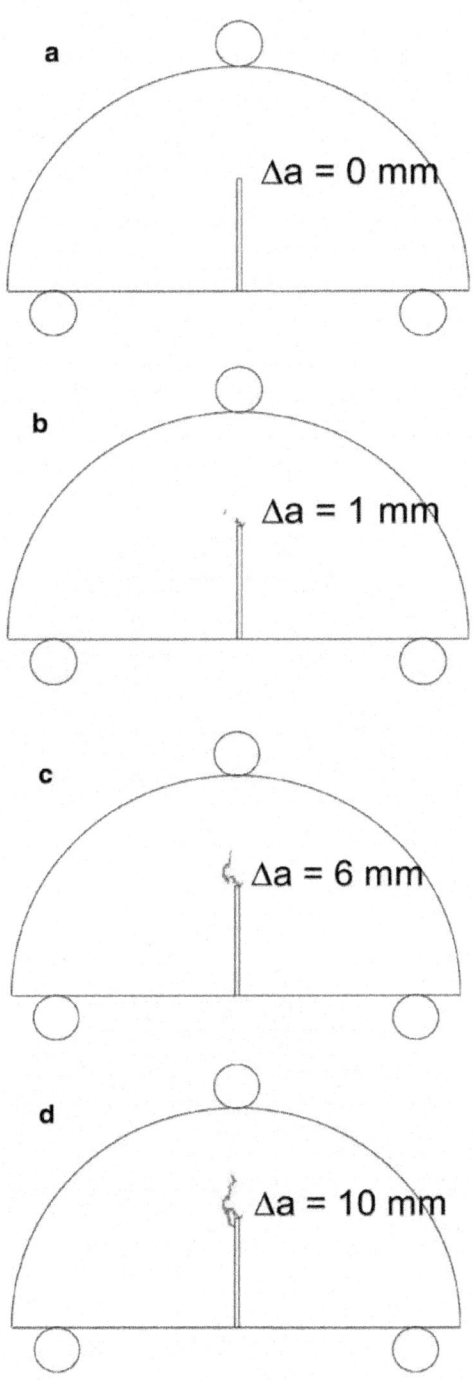

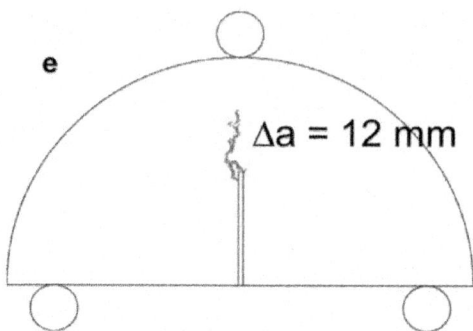

Figure. 12: Numerically determined crack growth during loading (the dots indicate the location of bond breakages) for a specimen with a/R = 0.5.

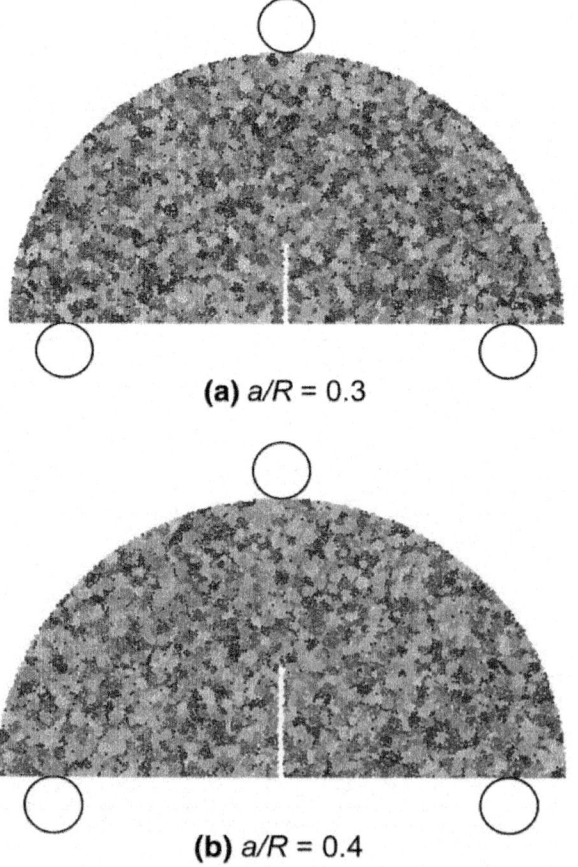

Figure. 13: SCB specimen model with (**a**) a/R = 0.3 and (**b**) a/R = 0.4.

Figure 14 shows the load and the number of bond breakages versus the load-line displacement, and Figure. 15shows the crack growths for those two cases. Table 4 gives a summary of the values of important parameters in the development of the stress intensity factor for a/R = 0.3, a/R = 0.4, and a/R = 0.5, respectively, and Figure. 16 shows the development of the stress intensity factor with crack growth, which is the K-resistance curve. These Figures show that the fracture toughness is about 0.95 MPam$^{0.5}$, which is comparable to the level II fracture toughness obtained for the CB specimen.

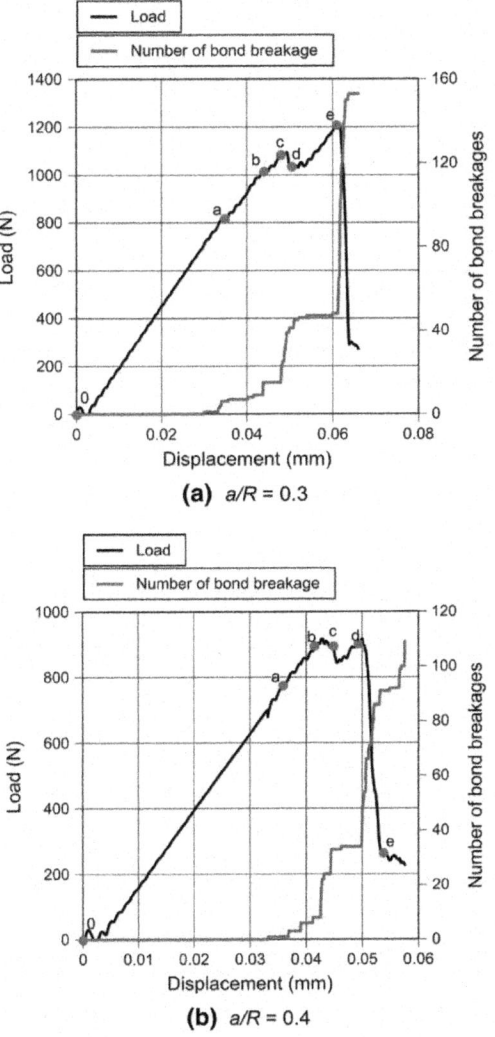

Figure. 14: Load versus displacement curves and cumulative number of bond breakages obtained by numerical simulation: **a** a/R = 0.3 and **b** a/R = 0.4.

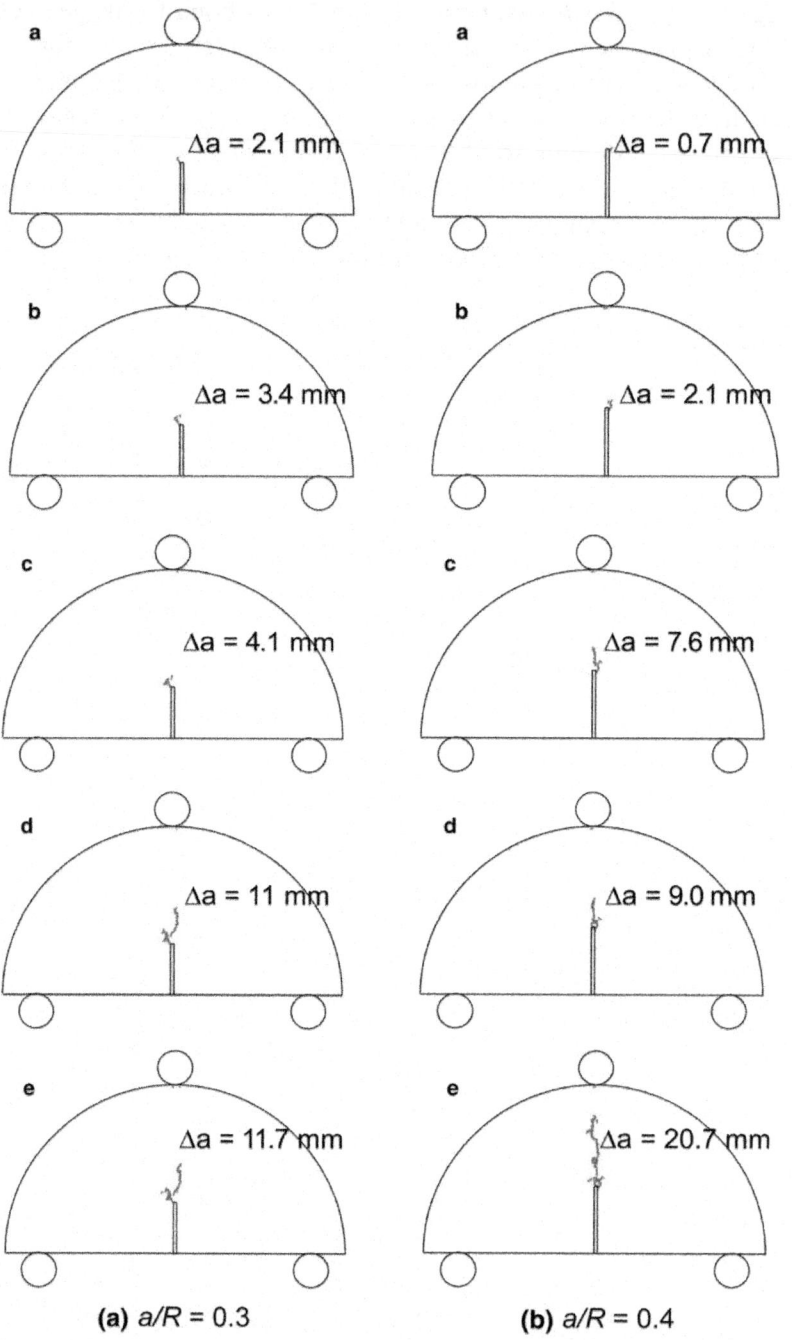

(a) *a/R* = 0.3   **(b)** *a/R* = 0.4

Figure. 15: Numerically determined crack growth during loading (the dots indicate the location of bond breakages) for a specimen with **a** a/R = 0.3 and **b** a/R = 0.4.

Table 4: Values used in the development of the stress intensity factor for (a) a/R = 0.3, (b) a/R = 0.4, and (c)a/R = 0.5

Part	R(mm)	S(mm)	t(mm)	a_0(mm)	Δa(mm)	a(mm)	a/R(−)	F (N)	σ_0(MPa)	Y (−)	K_{SCB}(MPa·m$^{0.5}$)
(a) a/R = 0.3											
0	50	40	25	15	0	15	0.3	0	0	4.36	0
a	50	40	25	15	2.1	17.1	0.34	806.3	0.323	4.55	0.34
b	50	40	25	15	3.4	18.4	0.37	1011.6	0.405	4.72	0.46
c	50	40	25	15	4.1	19.1	0.38	1077.8	0.431	4.84	0.51
d	50	40	25	15	11.0	26.0	0.52	1036.9	0.415	6.59	0.78
e	50	40	25	15	11.7	26.7	0.53	1211.4	0.485	6.84	0.96
(b) a/R = 0.4											
0	50	40	25	20	0	20	0.4	0	0	5.00	0
a	50	40	25	20	0.7	21	0.41	793.7	0.317	5.14	0.42
b	50	40	25	20	2.1	22	0.44	906.5	0.363	5.45	0.52
c	50	40	25	20	7.6	28	0.55	906.5	0.363	7.19	0.77
d	50	40	25	20	9.0	29	0.58	911.9	0.365	7.79	0.86
e	50	40	25	20	20.7	41	0.81	286.1	0.114	23.52	0.96
(c) a/R = 0.5											
0	50	40	25	25	0	25	0.5	0	0	6.26	0
a	50	40	25	25	0.0	25.0	0.5	474.2	0.190	6.26	0.33
b	50	40	25	25	1.0	26.0	0.52	616.7	0.247	6.59	0.46
c	50	40	25	25	6.0	31.0	0.62	744.1	0.298	8.91	0.83
d	50	40	25	25	10.0	35.0	0.7	548.5	0.219	12.38	0.90
e	50	40	25	25	12.0	37.0	0.74	480.0	0.192	15.18	1.00

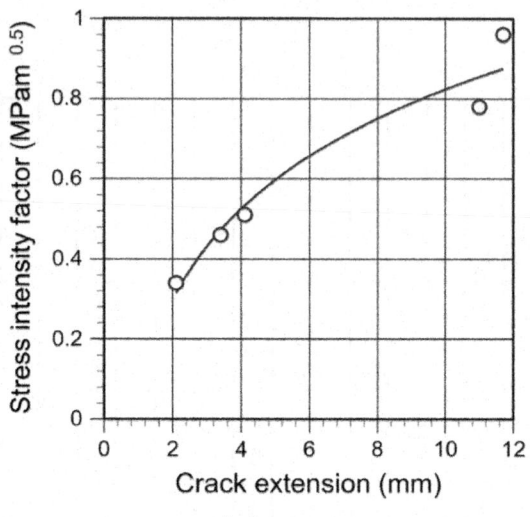

(a) *a/R* = 0.3

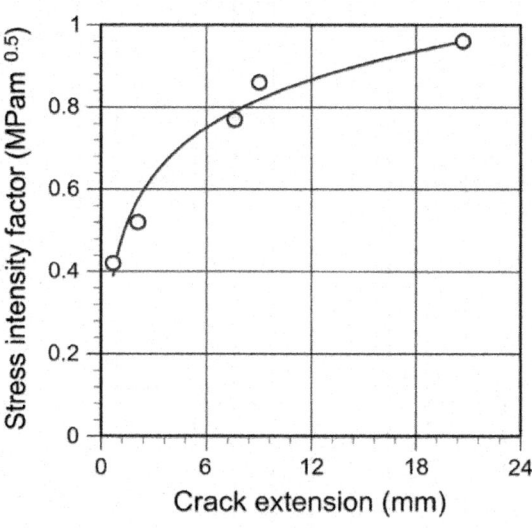

(b) *a/R* = 0.4

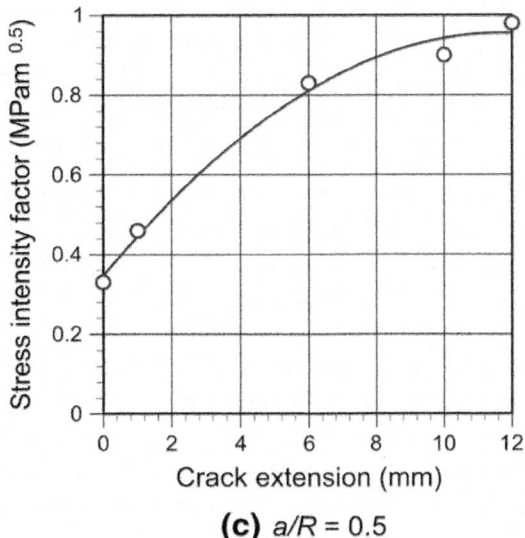

(c) *a/R* = 0.5

Figure. 16: K-resistance curves for Kimachi sandstone for: **a** a/R = 0.3, **b** a/R = 0.4, and **c** a/R = 0.5

The fundamental process of macrocrack extension in brittle rock is almost always by the opening, growth, and coalescence of microcracks that occurs within the process zone. Barker's nonlinearity correction incorporates the effect of the finite size of the process zone and yields an improved value known as the level II fracture toughness (Barker 1977). On the other hand, the K-resistance curve indirectly measures the energy release rate with the development of the process zone and, therefore, the two methods are expected to result in the same value for fracture toughness.

CONCLUSIONS

To investigate the mode I fracture toughness using semi-circular bend (SCB) specimens, we experimentally studied the fracture toughness using SCB and chevron bend (CB) specimens, the latter being one of the International Society for Rock Mechanics (ISRM) suggested methods, for comparison. The mode I fracture toughness measured using SCB specimens is lower than the level I and level II fracture toughness values measured using CB specimens. This difference is attributed to differences in notch type, neglecting the stable crack extension, and the differences in the fracture process zone (FPZ) sizes.

A numerical study based on discontinuum mechanics was conducted to evaluate crack propagation in the SCB specimen during loading. The numerical result is validated by comparing the load versus crack-opening displacement

(COD) curves obtained by the numerical model and by the experiments. The results show that subcritical crack growth, as well as sudden crack propagation, occurs when the load reaches the maximum. For the specimen sizes used for the tests, the crack extension at the maximum load is less than ~7 mm. This is almost the same as the radius of the FPZ as calculated by Schmidt's formula.

Moreover, the K-resistance curve is determined using the crack extension and the stress intensity factor at the evaluation point. The resistance increases with crack growth and reaches a steady value, which is considered to be the fracture toughness. The fracture toughness evaluated in this matter is in agreement with the level II fracture toughness measured using CB specimens. Therefore, the results show that an improved value for fracture toughness can be determined when the resistance to crack propagation is considered. The suggested numerical method enables the fracture toughness (which is comparable to that given by the ISRM suggested CB specimen method) to be determined by measuring the K-resistance during stable crack propagation in SCB specimens.

List of Symbols

a: Notch length

a_0: Chevron tip distance from the specimen surface

A_3: Coefficient of the higher-order non-singular term in the crack tip stress function

A_{min}: Minimum normalized stress intensity factor for the CB specimen

BDT: Uncracked Brazilian disk test

CB: Chevron bend

CCNBD: Cracked chevron-notched Brazilian disk

COD: Crack-opening displacement

CSTBD: Cracked straight-through Brazilian disk

D: Diameter of specimen

DEM: Distinct element method

E_{50}: Tangent Young's modulus at half the compressive strength

F: Load

F_c: Load at the evaluation point for the CB specimen

F_{max}: Maximum load

FEM: Finite element method

FPZ: Fracture process zone

K_{Ic}: Mode I fracture toughness

K_{CB}: Level I fracture toughness measured using the CB specimen

$K_{CB}{}^c$: Level II fracture toughness measured using the CB specimen

K_{SR}: Level I fracture toughness measured using the SR specimen

$K_{SR}{}^c$: Level II fracture toughness measured using the SR specimen

LEFM: Linear elastic fracture mechanics

LPD: Load-point displacement

MTS: Maximum tangential stress

p: Degree of nonlinearity

R: SCB specimen radius

r_c: Radius of the FPZ

t: SCB specimen thickness

S: Half span length of support rollers

SCB: Semi-circular bend

SECRBB: Single edge cracked round bar bend

SNDB: Straight notched disk bending

SR: Short rod

UCS: Uniaxial compressive strength

XRD: X-ray diffractometry

Y: Normalized stress intensity factor

σ_c: Uniaxial compressive strength

σ_t: Tensile strength

Δa: Crack extension

β: Ratio of notch length to specimen radius

v: Poisson's ratio

ACKNOWLEDGMENTS

We would like to thank Ms. Qian Li for her help with running the simulations.

REFERENCES

1.   Aliha MRM, Sistaninia M, Smith DJ, Pavier MJ, Ayatollahi (2012) Geometry effects and statistical analysis of mode I fracture in guiting limestone. Int J Rock Mech Min Sci 51:128–135. doi:10.1016/j.

ijrmms.2012.01.017

2. Azevedo NM, Lemos JV (2006) Hybrid discrete element/finite element method for fracture analysis. Comput Methods Appl Mech Eng 195(33–36):4579–4593. doi:10.1016/j.cma.2005.10.005

3. Barker LM (1977) A simplified method for measuring plane strain fracture toughness. Eng Fract Mech 9:361–369

4. Chang SH, Lee CI, Jeon S (2002) Measurement of rock fracture toughness under modes I and II and mixed-mode conditions by using disc-type specimens. Eng Geol 66(1–2):79–97. doi:10.1016/s0013-7952(02)00033-9

5. Cho N, Martin CD, Sego DC (2007) A clumped particle model for rock. Int J Rock Mech Min Sci 44(7):997–1010. doi:10.1016/j.ijrmms.2007.02.002

6. Chong KP, Kuruppu MD (1988) New specimens for mixed mode fracture investigations of geomaterials. Eng Fract Mech 30(5):701–712. doi:10.1016/0013-7944(88)90160-9

7. Chong KP, Kuruppu MD, Kuszmaul JS (1987) Fracture toughness determination of layered materials. Eng Fract Mech 28(1):43–54

8. Cundall PA, Strack OD (1979) A discrete numerical model for granular assemblies. Geotech 29(1):47–65

9. D'Addetta GA, Kun F, Ramm E (2002) On the application of a discrete model to the fracture process of cohesive granular materials. Granul Matter 4(2):77–90. doi:10.1007/s10035-002-0103-9

10. Dai F, Xia K, Nasseri MHB, Mohanty B (2007) Crack microcrack interaction and fracture toughness anisotropy in rocks. In: Proceedings of the SEM annual conference and exposition on experimental and applied mechanics, Springfield, MA, USA, June 2007, pp 1383–1390

11. Fowell RJ (1995) Suggested method for determining mode I fracture toughness using cracked chevron notched Brazilian disc (CCNBD) specimens. Int J Rock Mech Min Sci Geomech Abstr 32(1):57–64. doi:10.1016/0148-9062(94)00015-U

12. Fowell RJ, Xu C (1994) The use of the cracked Brazilian disc geometry for rock fracture investigations. Int J Rock Mech Min Sci Geomech Abstr 31(6):571–579

13. Funatsu T, Seto M, Shimada H, Matsui K, Kuruppu M (2004) Combined effects of increasing temperature and confining pressure on the fracture toughness of clay bearing rocks. Int J Rock Mech Min Sci 41(6):927–938. doi:10.1016/j.ijrmms.2004.02.008

14. Funatsu T, Li Q, Shimizu N, Seto M, Matsui K (2008) Numerical

simulation of crack propagation in rock by particle flow code. J MMIJ 124(10–11):611–618

15. Guo H, Aziz NI, Schmidt LC (1993) Rock fracture-toughness determination by the Brazilian test. Eng Geol 33(3):177–188. doi:10.1016/0013-7952(93)90056-I

16. International Society for Rock Mechanics (ISRM) (1988) Suggested methods for determining the fracture toughness of rock. Int J Rock Mech Min Sci Geomech Abstr 25(2):71–96. doi:10.1016/0148-9062(88)91871-2

17. Iqbal MJ, Mohanty B (2007) Experimental calibration of ISRM suggested fracture toughness measurement techniques in selected brittle rocks. Rock Mech Rock Eng 40(5):453–475. doi:10.1007/s00603-006-0107-6

18. Itasca Consulting Group Inc. (2004) Particle flow code in 2-dimensions (PFC2D) version 3.10, Minneapolis, MN

19. Khan K, Al-Shayea NA (2000) Effect of specimen geometry and testing method on mixed mode I–II fracture toughness of a limestone rock from Saudi Arabia. Rock Mech Rock Eng 33(3):179–206. doi:10.1007/s006030070006

20. Kuruppu MD, Obara Y, Ayatollahi MR, Chong KP, Funatsu T (2014) ISRM-suggested method for determining the mode I static fracture toughness using semi-circular bend specimen. Rock Mech Rock Eng 47(1):267–274. doi:10.1007/s00603-013-0422-7

21. Lim IL, Johnston IW, Choi SK, Boland JN (1994) Fracture testing of a soft rock with semi-circular specimens under three-point bending. Part 1—mode I. Int J Rock Mech Min Sci Geomech Abstr 31(3):185–197. doi:10.1016/0148-9062(94)90463-4

22. Matsuki K, Hasibuan SS, Takahashi H (1991) Specimen size requirements for determining the inherent fracture toughness of rocks according to the ISRM suggested methods. Int J Rock Mech Min Sci Geomech Abstr 28(5):365–374. doi:10.1016/0148-9062(91)90075-W

23. Nasseri MHB, Mohanty B, Young RP (2006) Fracture toughness measurements and acoustic emission activity in brittle rocks. Pure Appl Geophys 163(5–6):917–945. doi:10.1007/s00024-006-0064-8

24. Ouchterlony F (1981) Extension of the compliance and stress intensity formulas for the single edge crack round bar in bending. ASTM STP 745:237–256

25. Potyondy DO, Cundall PA (2004) A bonded-particle model for rock. Int J Rock Mech Min Sci 41(8):1329–1364. doi:10.1016/j.ijrmms.2004.09.011

26. Schmidt RA (1980) A microcrack model and its significance to hydraulic fracturing and fracture toughness testing. In: Proceedings of the 21st U.S. symposium on rock mechanics (USRMS), Rolla, MO, USA, May 1980, pp 581–590

27. Tan Y, Yang D, Sheng Y (2009) Discrete element method (DEM) modeling of fracture and damage in the machining process of polycrystalline SiC. J Eur Ceram Soc 29(6):1029–1037. doi:10.1016/j.jeurceramsoc.2008.07.060

28. Tutluoglu L, Keles C (2011) Mode I fracture toughness determination with straight notched disk bending method. Int J Rock Mech Min Sci 48(8):1248–1261. doi:10.1016/j.ijrmms.2011.09.019

29. Wang Q-Z, Xing L (1999) Determination of fracture toughness K_{IC} by using the flattened Brazilian disk specimen for rocks. Eng Fract Mech 64:193–201

Chapter 7

ESTIMATING HYDRAULIC CONDUCTIVITY OF HIGHLY DISTURBED CLASTIC ROCKS IN TAIWAN

Cheng-Yu Ku[1] and Shih-Meng Hsu[2]

[1]National Taiwan Ocean University
[2]Sinotech Engineering Consultants, Inc Taiwan

INTRODUCTION

Understanding groundwater flow in fractured consolidated media has long been important when undertaking engineering tasks such as dam construction, mine development, the abstraction of petroleum, slope stabilization, and the construction of foundations. To study groundwater flow in support of these tasks, the focus of most hydrogeological investigations has been on the characterization of the hydraulic properties of the higherpermeability fractures in the rock mass.

Taiwan is situated on the edge of the Eurasian and Philippine Sea plate. Plate tectonics have created numerous fault lines that crisscross the island. As a result of high density of faults, rock core data with fractures, soft and cohesive gouges, and various lithologies are extensive in boreholes. In general, the permeability of clay-rich gouges has extremely low values. On the contrary, the fractures often have higher permeability. The hydraulic properties of fractured rocks in Taiwan, therefore, vary with highly disturbed geological structures and lithology. To obtain hydraulic properties of fractured rocks in Taiwan, the investigation of vertical variation of the fractures in a borehole is of importance. This study utilized a highresolution BoreHole acoustic TeleViewer (BHTV, Williams and Johnson, 2004) to scan images of the borehole. The information gathered from BHTV was used to characterize lithology and fractures for the borehole and was essential to conduct a proper measurement of rock mass hydraulic conductivity. The double packer systems were then used to determine the hydraulic conductivity in a portion of borehole using two inflatable packers. Although this type of test can directly measure the hydraulic parameter, costs of the testing are fairly high. Several studies (Black, 1987; Carlson and Olsson,1977; Louis, 1974; Burgess, 1977; Wei et al., 1995, Zhao, 1998) have proposed the estimation of rock mass hydraulic conductivity using different empirical equations. These empirical equations provide a great feature for characterizing rock mass hydraulic properties quickly and easily. However, the applicability of these equations in highly disturbed clastic sedimentary rocks in Taiwan is very limited.

This study proposed the establishment of an empirical HC model for estimating rock mass hydraulic conductivity of highly disturbed clastic sedimentary rocks in Taiwan using the BHTV and the double packer hydraulic tests. Four geological parameters including rock quality designation (RQD), depth index (DI), gouge content designation (GCD), and lithology permeability index (LPI) were adopted for establishing the empirical HC model. To verify rationality of the proposed HC model, 22 in-situ hydraulic tests were carried out to measure the hydraulic conductivity of the highly disturbed clastic sedimentary rocks in three boreholes at two different locations in Taiwan. Besides, the model verification using another borehole data with four additional in-situ hydraulic tests from similar clastic sedimentary rocks was also conducted to further verify the feasibility of the proposed empirical HC model. This paper presents the measured hydraulic conductivity results and the relationship among the hydraulic conductivity, RQD, DI, GCD, and LPI. The application of the proposed HC model was also addressed.

DESCRIPTION OF STUDY AREAS AND BOREHOLES

Description of Study Areas

Taiwan's strata are distributed in long and narrow strips, almost parallel to the island's axis. Metamorphic rock lies under the Central and Snow Mountain Ranges. Sedimentary rock forms part of the island-wide piedmonts and coastal plains as well as the Coastal Mountain Range. The island of Taiwan has three geological zones divided by longitudinal faults: the Central Range, Western Piedmont and Eastern Coastal Mountain Range zones (Fig. 1(a)). About 26 hydraulic conductivity measurements were conducted in four boreholes in Western Piedmont, primarily at three sites: Da-Keng, Shang-Ming, and Caoling (Fig. 1(a)) in which borehole HB-94-01 is in the Da-Keng site, boreholes HB-95-01 and HB-95-02 are in the Shang-Ming site, and borehole CH-04 is in the Caoling site. Besides, the Da-Keng and Caoling sites are in central Taiwan and the Shang-Ming site is in southern Taiwan. The dominant rock strata of the

Shang-Ming site include Miocene sedimentary rock with layers of sandstone or shale or their alternation. The major structures consist of a series of parallel easterly inclined thrust faults and folds, which often form local fractured zones, including geological structures such as the Pingshi fault, the Biauhu fault, and the Chin-Shan fault. Figure 1(b) presents the distribution of these geological strata and structures. In addition, borehole HB-94-01 in the Da-Keng site and borehole CH-04 in the Caoling site also have similar rock strata but without geological structures.

Based on the loggings and geological analysis, HB-95-01 and HB-95-02 are strongly influenced by the faults; nevertheless, HB-94-01 and CH-04 are not.

Boreholes

The depth of the borehole HB-94-01 is 110 m. The principal lithologic units for the borehole are sandstone and siltstone. The interval of 36 m to 44 m is a fractured zone compared to other depths in the borehole. A total of 8 hydraulic tests using a double packer system were carried out to determine hydraulic conductivity (Sinotech, 2006). The strategy of the test design from the drilling work was to determine hydraulic properties from different geological structures such as no fracture, a single fracture, or multiple fractures at different depths.

The drilling depths of HB-95-01 and HB-95-02 are 250 m and 350 m, respectively. The principal lithologic units for HB-95-01 are sandstone, argillaceous sandstone, and sandy mudstone. The principal lithologic units for HB-95-02 are sandstone, argillaceous sandstone, and sandstone mixed with some mudstone. HB-95-01 and HB-95-02 are close to the Biauhu fault and Pingshi fault, respectively (Fig. 1(b)). Rock core photos (Fig. 2(a)) indicated soft and cohesive gouges are extensive in both boreholes in which the hydraulic properties of fault-related rocks can be studied. The study completed 3 and 14 hydraulic tests in HB-95-01 and HB-95-02, respectively (Sinotech, 2006). The strategy of the test design was to determine hydraulic conductivity in more permeable zones and clay-rich gouge zones. Besides, the borehole CH-04 is not influenced by the faults (Fig. 2(b)) and used for the model verification and it is described in Section 5.7.

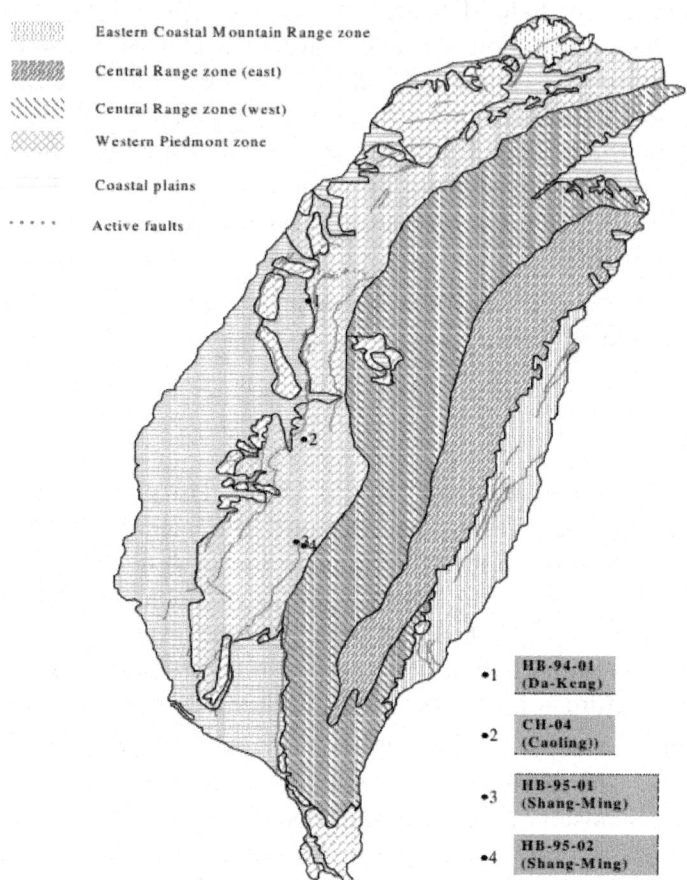

Figure 1. (a) Location of major faults and four boreholes for this study in Taiwan.

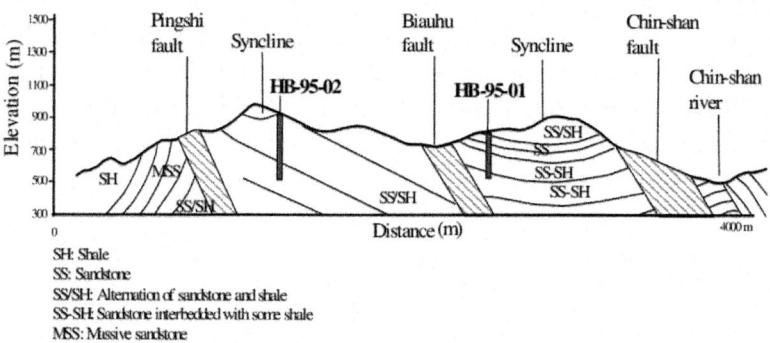

Figure 1. (b) Detailed distribution of geological strata and structures of boreholes HB-95-01 and HB-95-02.

Figure 2. (a) Rock core photos of borehole HB-95-2 with fault influence.

Figure 2. (b) Rock core photos of borehole CH-04 without fault influence.

HYDRAULIC CONDUCTIVITY OF FRACTURED ROCK MASSES

It is widely recognized that fracturing plays a decisive role in rock hydraulics, especially in low permeability rocks such as crystalline, volcanic and carbonate rocks, and in some classic sedimentary formations, such as sandstones, shales, glacial tills and clays. In highly disturbed fractured rocks, hydraulic properties depend on density, size, infillings and interconnection of fractures. A distinction can be made between hydraulic conductivity of fracture and of intergranular (matrix) material. Previous study (Lee and Farmer, 1993) has proposed the hydraulic conductivity of a rock mass with three orthogonal joint sets with similar spacing ad constant aperture in all directions. The effect of stress on permeability is also of importance for estimating rock mass hydraulic conductivity (Snow, 1969). Several studies, shown in Table 1, have also pointed out that rock mass permeability may decrease systematically with depth (Black, 1987; Carlson and Olsson,1977; Louis, 1974; Burgess, 1977; Wei et al., 1995, Zhao, 1998). The decrease in permeability with depth in fractured rocks is usually attributed to reduction in fracture aperture and fracture spacing. The reduction is due to the effect of geostatic stresses, and thereby the permeability of fractured rocks will be reduced. Accordingly, the depth may be considered as a factor in evaluating rock mass permeability.

MEASUREMENT OF ROCK MASS HYDRAULIC CONDUCTIVITY

For decades, the determination of hydraulic properties in fracture rocks has been qualitatively estimated using the Lugeon test. It is now recognized that this approach is not suitable in highly disturbed fractured rocks. The type of test only gives an average value of hydraulic conductivity in a stratum and is not able to identify (1) aquifer's type in a required testing section; (2) storativity of an aquifer; and (3) relations between hydraulic properties and geological structures such as water-bearing fractured zones. Results from the test may be insufficient to characterize hydraulic properties for complex geological environments. They may be subject to hazards such as extensive water inflow during underground excavation.

To provide a better characterization of hydraulic properties of fractured rocks, a double packer technique can be adopted and is often utilized to overcome the shortcomings of the Lugeon test. Packers can be used to isolate a portion of borehole for hydraulic testing. Hydraulic properties for a single of fracture, a group of fractures, or an entire rock formation can be easily identified by the technique.

Table 1. Diverse approximations for estimating rock mass hydraulic conductivity.

Equation	Reference
$k = az^{-b}$	Black (1987) a and b are constants, z is the vertical depth below the groundwater surface.
$\log K = -8.9 - 1.671 \log Z$	Snow (1969) K (ft²) is the permeability. z (ft) is the depth.
$K = 10^{-(1.6 \log z + 4)}$	Carlson and Olsson (1977) K (m/s) is the hydraulic conductivity. z (m) is the depth.
$K = K_s e^{(-Ah)}$	Louis (1974) K (m/s) is the hydraulic conductivity. K_s is the hydraulic conductivity near ground surface. h (m) is the depth. A is the hydraulic gradient.
$\log K = 5.57 + 0.352 \log Z$ $-0.978(\log Z)^2 + 0.167(\log Z)^3$	Burgess (1977) K (m/s) is the hydraulic conductivity. Z (m) is the depth.
$K = K_i[1 - Z / (58.0 + 1.02Z)]^3$	Wei et al. (1995) Z is the depth. K is the hydraulic conductivity. K_s (m/s) is the hydraulic conductivity near ground surface.

Borehole Acoustic Televiewer(BHTV) Investigation And Hydraulic Test Design

Prior to hydraulic testing, the study utilized a high-resolution borehole acoustic televiewer (BHTV) to scan images of boreholes. The information (Fig. 3(a) and (b)) gathered from BHTV was used to characterize lithology and fractures for the borehole and was essential to the proper design of the hydrogeological program. Test design is dependent on the characteristics of the zone tested and the desired information. Accordingly, the main testing strategy in this study was to detect waterbearing fractures. In addition, the study investigated the vertical variation of the hydraulic conductivity in a borehole and hydraulic property of fault-related rocks.

A water-bearing zone of subsurface commonly appears in the section with multiple fractures. According to BHTV logs from boreholes, the study selected locations with images that show multiple fractures as hydraulic test sections. Figure 3 shows that two test zones were selected by this strategy. Other testing zones for other study purposes can be selected by BHTV images.

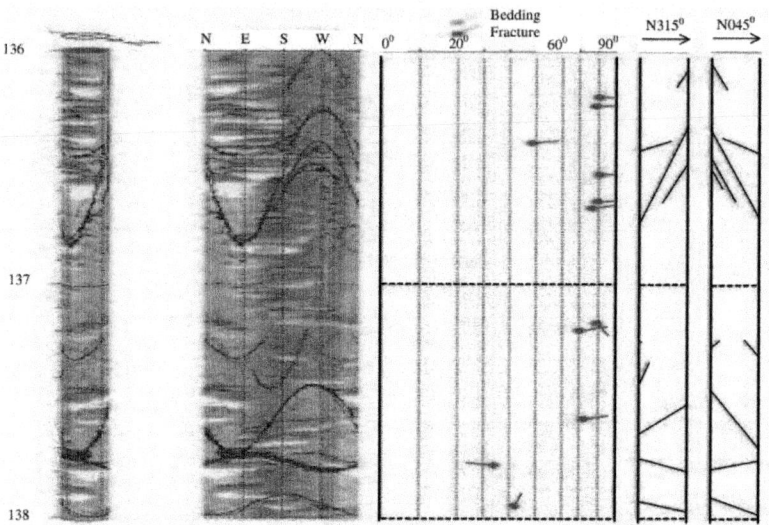

Figure 3. (a) The pack-off zones and their corresponding BHTV images (depth 136m~138m, HB-95-01).

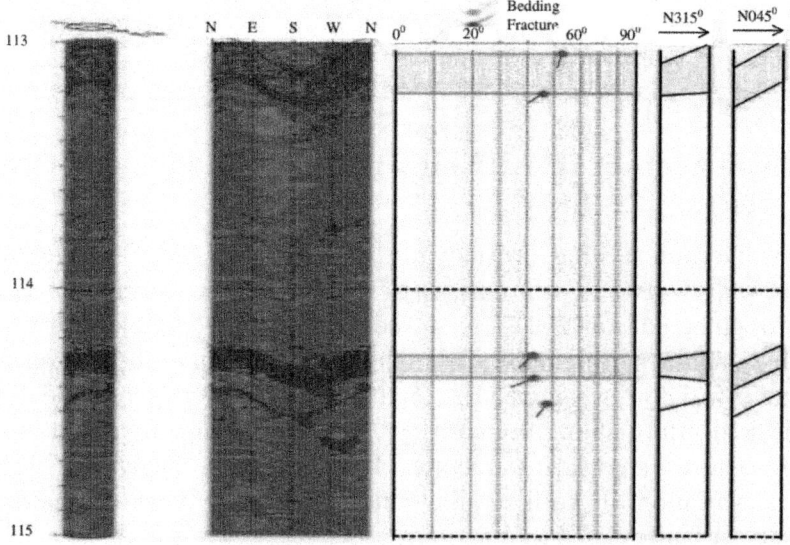

Figure 3. (b) Identification of shear band from BHTV (depth 113m~115m, HB-95-02).

Double Packer System

Double packer systems are the most commonly used tools for hydrogeological testing in boreholes. They can be used to determine the hydraulic property in a section of borehole based on two inflatable packers. It is now recognized that this approach is appropriate to investigate the variability of a borehole as it intersects various hydrogeological units. The double packer system in this study (Fig. 4) consists of two inflatable rubber packers, a shut-in valve, a submersible pump, and pressure transducers for monitoring above and below the packers and the isolated interval. The shut-in valve can open or close the hydraulic connection between the pipe string and the test section. The rubber packers can be inflated using nitrogen delivered through a polyethylene air line. The pumping or injecting rate can be monitored at the land surface with a flow meter. To conduct each hydraulic test, the packers are inflated to isolate a section of borehole, and the rate of flow and/or pressure in the test interval over a period of time can be measured.

The BHTV images for different test intervals in Boreholes HB-95-01 and HB-95-02 are shown in Fig. 5(a) and 5(b). It is obvious that the fractures can be identified clearly using our highresolution BHTV. The intervals in the depth from 118.5 m to 121.7 m and from 134.8 m to 138.0 m were sealed by double packers for conducting a pressure pulse test and a constant head injection test, respectively. Figures 5(a) and 5(b) show the results of hydraulic tests which were conducted by different type of hydraulic tests by means of AQTESOLVE. The type of hydraulic test chosen in this study for each test interval was decided by a hydraulic diagnosis test which mainly detects permeability of the test interval prior to a normal test. For the test interval of 118.5 m to 121.7 m, although three fractures and a fracture zone of approximately 7.25 cm thickness were seen on the borehole image, lack of interconnectivity of fractures and soft and cohesive gouges existing at the fractures may reduce the permeability of rock masses. In addition, four types of tests, including pumping tests, injection tests, slug tests, and pressure pulse tests can be applied to the double packer system. Pumping tests involve pumping at a constant or variable rate and measuring changes in water levels during pumping. In injection tests, fluid is injected into a test interval while keeping the head of the test interval at a constant value. A slug test involves the abrupt removal, addition, or displacement of a known volume of water and the subsequent monitoring of changes in water level as equilibrium conditions return. In a pressure pulse test, an increment of pressure is applied to a packed zone. The pressure decay is monitored. Typically, the decision on which type of test to perform is based on the expected permeability of the test interval, the volume of rock to be sampled, and the availability of time and equipment (NRC, 1996). Hydraulic properties determined by slug tests or pressure pulse tests are representative only for the material in the immediate vicinity of the borehole.

To obtain hydraulic conductivity over a large area, the procedure of a single-hole hydraulic test is to perform a pumping test at a test interval first. If the pumping test cannot be performed due to low permeability of the test section, a constant head injection test will be conducted instead. Once the flow rate cannot be measured by limitation of the flow meter (less than 0.11 l/min) during the injection test, a slug test or pressure pulse test can be performed. The duration of a pressure pulse test is much shorter than that of a slug test. For this reason, the pressure pulse test is commonly applied to test intervals of very low permeability. However, the volume of rock tested by a pressure pulse test is significantly smaller when compared to a slug test.

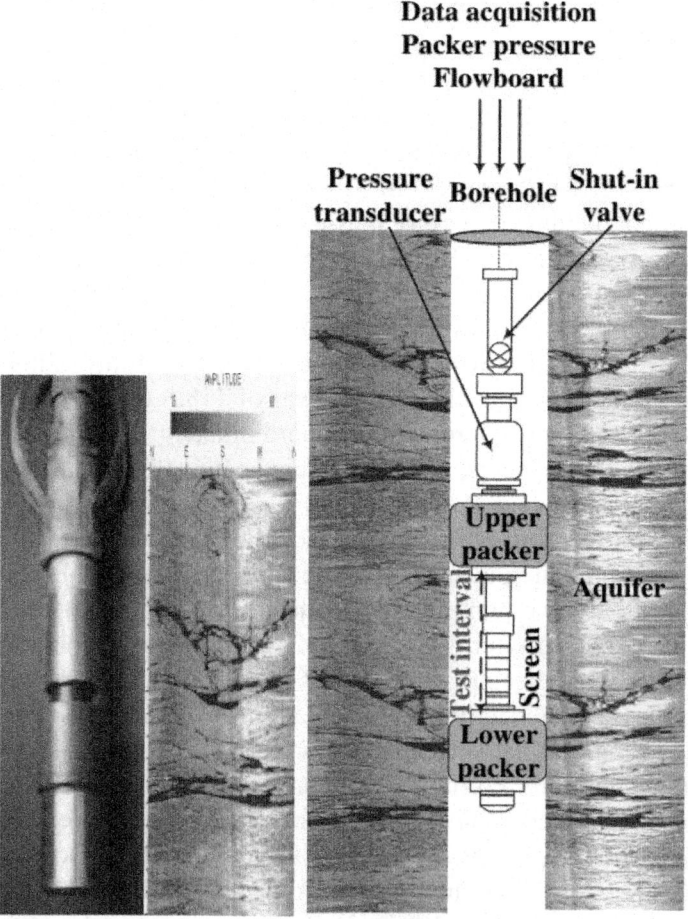

Figure 4. Schematic drawing of BHTV, acoustic image of borehole, and the double packer system.

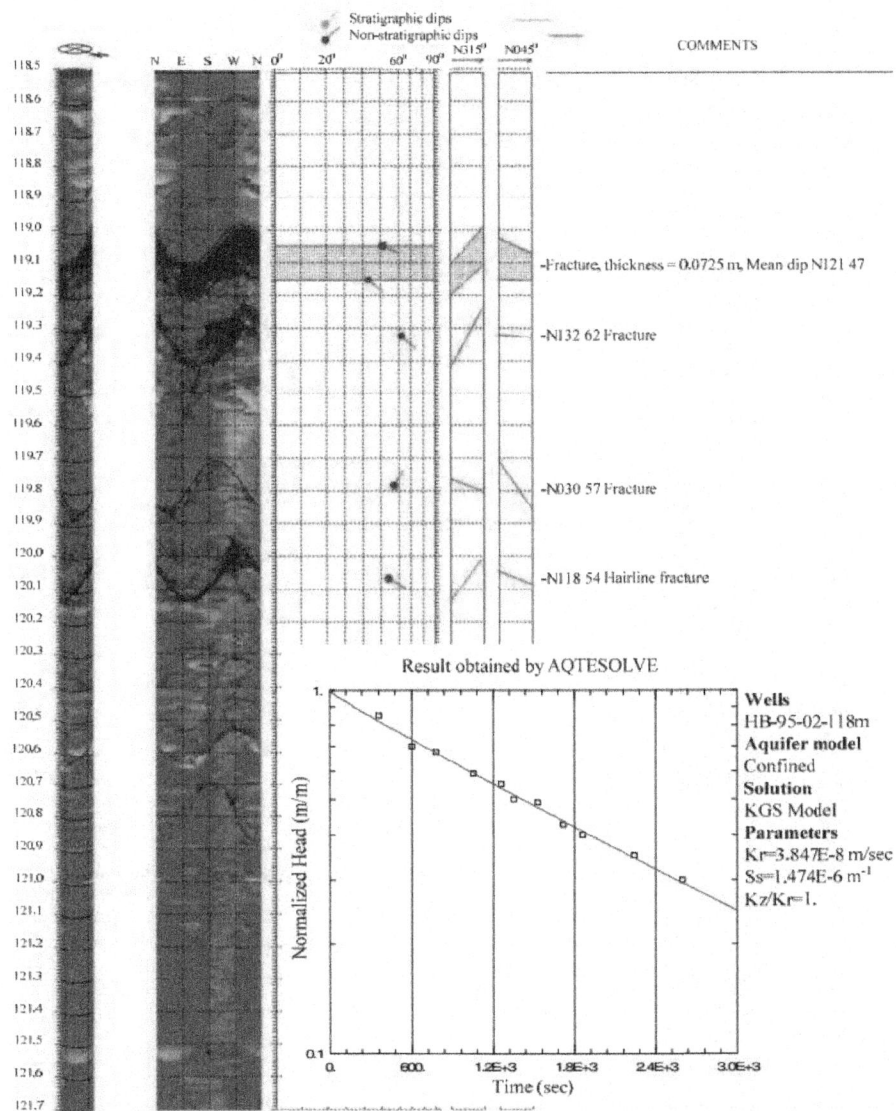

Figure 5. (a) Evaluation of hydraulic parameters using AQTESOLVE (right lower figure) and BHTV images at pack-off zones 118.5 m to 121.7 m in Borehole HB-95-02.

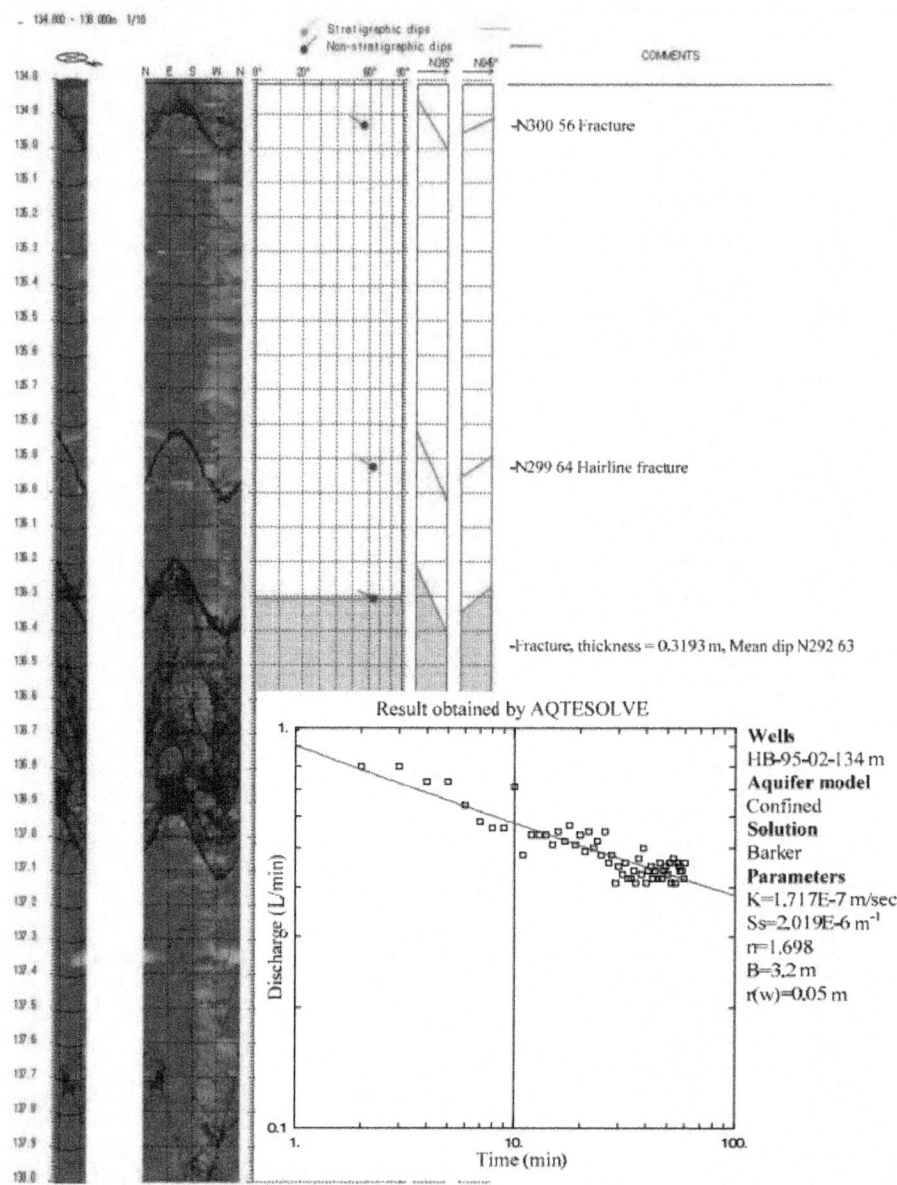

Figure 5. (b) Evaluation of hydraulic parameters using AQTESOLVE (right lower figure) and BHTV images at pack-off zones 134.8 m to 138 m in Borehole HB-95-02.

A total of 26 hydraulic tests were designed to determine hydraulic conductivity in the boreholes. Data collected during hydraulic tests can be analyzed by analytical methods. Water pressure and discharge rate measurements with time for each hydraulic test were collected in this study. The data analysis was performed using a professional version of the AQTESOLVE test analysis software, which enables both virtual and automatic type curve matching (Duffield, 2004). The quantitative evaluation of hydraulic parameters was carried out as an iterative process of the best-fit theoretical response curves based on the measured data of the hydraulic test.

EMPIRICAL MODEL OF ROCK MASS HYDRAULIC CONDUCTIVITY

Prior to describing the empirical model of rock mass hydraulic conductivity, an attempt to find the decrease in permeability with depth was conducted. Figure 6 demonstrates that the testing data of HB-94-01 shows the tendency that the hydraulic conductivity decreases with depth. The form of the regression equation is close to the result obtained by Black, 1987. The coefficient of determination of the regression equation is 0.633. However, the testing data from HB-95-01 and HB-95-02 are very scattered. No relationship can be found between hydraulic conductivity and depth. Accordingly, potential factors, including rock quality designation (RQD), depth index (DI), gouge content designation (GCD), and lithology permeability index (LPI), that may affect the degree of permeability should be considered. The rating approach for each factor that represents the magnitude of permeability is also described as following.

Rock Quality Designation

To assess the influence of the fracture characteristic on permeability, the rock quality designation (RQD) index (Deere et al., 1967), can be adopted. The RQD index was introduced over 40 years ago as an indicator of rock mass conditions. The RQD value is defined as the cumulative length of core pieces longer than 100 mm in a run (RS) divided by the total length of the core run (RT) and can be obtained from the following equation.

$$\text{RQD} = \frac{\sum \text{Length of Intact and Sound Core Pieces} > 100 \text{ mm}}{\text{Total Length of Core Run, mm}} \times 100\%$$

$$= \frac{R_S}{R_T} \times 100\% \tag{1}$$

In this study, a core run for calculating a RQD value is herein defined as a selected zone of a hydraulic test. Eq. 1. may be utilized to identify rock mass permeability.

Depth Index

The decrease in permeability with depth in fractured rocks is usually attributed to reduction in fracture aperture and fracture spacing. The reduction is due to the effect of geostatic stresses, and thereby the permeability of fractured rocks will be reduced. The depth may be considered as a factor in evaluating rock mass permeability. To assess the influence of the depth on permeability, a Depth Index, namely DI, was defined as the following equation.

$$DI = 1 - \frac{L_c}{L_T}$$

(2)

in which LT is the total length of a borehole; LC is a depth which is located at the middle of a double packer test interval in the borehole. The value of DI is always greater than zero and less than one. The greater the DI value, the higher the permeability.

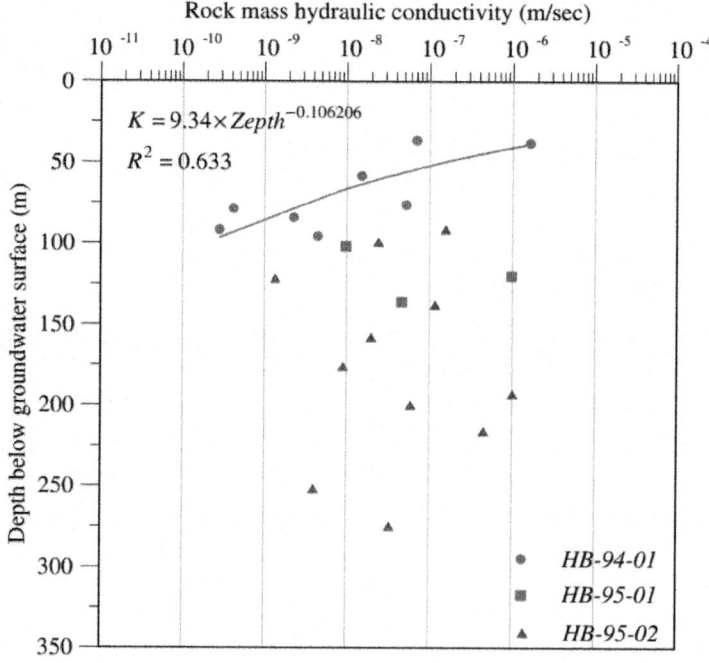

Figure 6. Relationship between hydraulic conductivity and depth.

Gouge Content Designation

The RQD value may decrease by an increase of fractures in a core run. If the fractures contain infillings such as gouges, permeability of the fractures will reduce. To assess the influence of the gouge materials on permeability, a Gouge Content Designation (GCD) index was defined as the following equation.

$$GCD = \frac{R_G}{R_T - R_S},$$

(3)

in which RG is the total length of gouge content. The value of GCD is always greater than zero and less than one. The greater GCD value stands for the more gouge content in a core run, and thereby it will reduce the permeability.

Lithology Permeability Index

Lithology is the individual character of a rock in terms of mineral composition, grain size, texture, color, and so forth. For an intact rock, the magnitude of permeability depends largely on the individual character of the rock. It may be affected by the average size of the pores, which in turn is related to the distribution of particle sizes and particle shape. In sedimentary formations grain-size characteristics are most important because coarsegrained and well-sorted material will have high permeability as compared to fine-grained sediments like silt and clay. Thus, the lithology may be regarded as a factor in evaluating rock mass permeability. To assess the influence of lithology on permeability, a Lithology Permeability Index (LPI) was defined as Table 2.

Table 2. Description and ratings for lithology permeability index.

Lithology	Hydraulic conductivity (m/s)				Range of rating	Suggested Rating
	Reference[1]	Reference[2]	Reference[3]	K$_{average}$		
Sandstone	10^{-6}~10^{-9}	10^{-7}~10^{-9}	10^{-7}~10^{-9}	$10^{-7.5}$	0.8-1.0	1.00
Silty Sandstone	–	–	–	–	0.9-1.0	0.95
Argillaceous Sandstone	–	–	–	–	0.8-0.9	0.85
S.S. interbedded with some Sh.	–	–	–	–	0.7-0.8	0.75
Alternations of S.S & Sh.	–	–	–	–	0.6-0.7	0.65
Sh. interbedded with some S.S.	–	–	–	–	0.5-0.7	0.60
Alternations of S.S & Mudstone	–	–	–	–	0.5-0.6	0.55
Dolomite	10^{-6}~$10^{-10.5}$	10^{-7}~$10^{-10.5}$	10^{-9}~10^{-10}	10^{-8}	0.6-0.8	0.70
Limestone	10^{-6}~$10^{-10.5}$	10^{-7}~10^{-9}	10^{-9}~10^{-10}	10^{-8}	0.6-0.8	0.70
Shale	10^{-10}~10^{-12}	10^{-10}~10^{-13}	–	$10^{-10.5}$	0.4-0.6	0.50
Sandy Shale	–	–	–	–	0.5-0.6	0.60
Siltstone	10^{-10}~10^{-12}	–	–	10^{-11}	0.2-0.4	0.30
Sandy Siltstone	–	–	–	–	0.3-0.4	0.40
Argillaceous Siltstone	–	–	–	–	0.2-0.3	0.20
Claystone	–	10^{-9}~10^{-13}	–	10^{-11}	0.2-0.4	0.30
Mudstone	–	–	–	–	0.2-0.4	0.20
Sandy Mudstone	–	–	–	–	0.3-0.4	0.40
Silty Mudstone	–	–	–	–	0.2-0.3	0.30
Granite	–	–	10^{-11}~10^{-12}	$10^{-11.5}$	0.1-0.2	0.15
Basalt	10^{-6}~$10^{-10.5}$	10^{-10}~10^{-13}	–	$10^{-11.5}$	0.1-0.2	0.15

[1]B.B.S. Singhal & R.P. Gupta (1999) ; [2]Karlheinz Spitz & Joanna Moreno (1996) ; [3]Bear(1972)

Rock Mass Permeability Index

As stated above, the rock mass permeability may be dependent on the following four parameters: RQD, DI, GCD, and LPI. However, the permeability is not simply affected by only one factor. It may account for the synthetic effect from the four parameters on permeability. Accordingly, Rock mass permeability index, called the HC index, was proposed herein.

$$HC = (1\text{-}RQD)(DI)(1\text{-}GCD)(LPI)\ .,$$

$$(4)$$

The value of each parenthesis at the right hand side of Eq. 4. is always greater than zero and less than one depending on the values assigned to the four parameters. The greater the value of each parenthesis, the higher the permeability. Thus, the model performs a numerical assessment of rock mass permeability using the four parameters. Since it is rare to encounter the condition that RQD is 100% in highly disturbed clastic sedimentary rocks in Taiwan, the term of (1-RQD) is usually greater than zero. However, it should be noted that if (1-RQD) is zero, the value of 0.01 in the term of (1-RQD) is suggested to avoid the HC value to be zero. Currently, the study took the same weight for each factor in Eq. (4). In addition, Eq. (4) is limited in sedimentary rocks only and is only applied to vertical boreholes at present. With more testing data, a further study can be considered to assign a different weight for each factor to give a better correlation between the hydraulic conductivity and HC.

The Empirical HC Model

Regression analysis was performed to estimate the dependence of HC on hydraulic conductivity. A total of 22 hydraulic test data were applied to the study. HC-values for the hydraulic tests can be computed from borehole image data and rock core data, in which the values of RQD and GCD at each test interval can be calculated from borehole image data and rock core data with Eqs. 1. and 3., respectively. The value of DI can be calculated using Eq. 2. The value of LPI for each test zone can be obtained from rock core data and Table 3. Table 3 shows the calculated results for the HC model based on the verified data. The regression results indicated that a power law relationship exists between the hydraulic conductivity and HC with a coefficient of determination of 0.866 as shown in Fig. 7. The empirical HC model is obtained as shown in Eq. 5.

$$K = 2.93 \times 10^{-6} \times (HC)^{1.380}, R^2 = 0.866$$

$$(5)$$

If only HB-94-01 testing data were adopted, a better correlation with the coefficient of determination of 0.905 can be obtained as shown in Eq. 6.

$$K = 2.31 \times 10^{-6} \times (\text{HC})^{1.342}, R^2 = 0.905 \qquad (6)$$

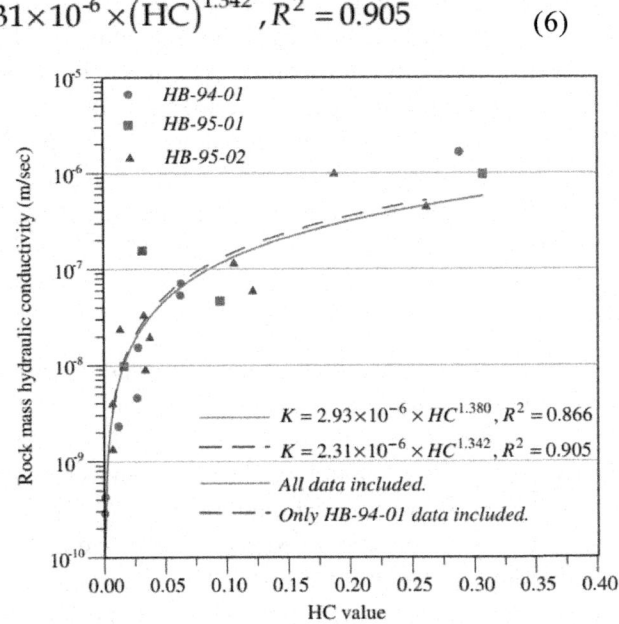

Figure 7. Relationship between hydraulic conductivity and HC-values.

Table 3. The calculated results for HC-system based on 22 hydraulic test data.

Boreholes	Test intervals (m)	1-RQD	DI	1-GCD	LPI	HC	K (m/s)
HB-94-01	34.7-36.3	0.094	0.677	1.000	1.000	0.0635	7.06E-08
	36.4-38.0	0.438	0.662	1.000	1.000	0.2895	1.64E-06
	56.7-58.3	0.063	0.477	1.000	0.950	0.0283	1.53E-08
	74.6-76.2	0.500	0.315	1.000	0.400	0.0629	5.3E-08
	77.2-78.8	0.010	0.291	1.000	0.400	0.0012	4.22E-10
	82.6-84.2	0.125	0.242	1.000	0.400	0.0121	2.31E-09
	90.2-91.8	0.010	0.173	1.000	0.400	0.0007	2.86E-10
	94.2-95.8	0.500	0.136	1.000	0.400	0.0273	4.53E-09
HB-95-01	99.0-101.9	0.345	0.598	0.200	0.400	0.0165	9.8E-09
	117.2-120.1	0.690	0.526	1.000	0.850	0.3081	9.76E-07
	133.2-136.1	0.724	0.461	0.286	1.000	0.0954	4.68E-08
HB-95-02	88.6-91.4	0.071	0.743	1.000	0.600	0.0318	1.56E-07
	96.0-99.2	0.031	0.721	1.000	0.600	0.0135	2.42E-08
	118.5-121.7	0.219	0.657	0.071	0.700	0.0072	1.36E-09
	134.8-138.0	0.344	0.610	0.727	0.700	0.1068	1.17E-07
	154.8-158.0	0.938	0.553	0.103	0.700	0.0376	1.99E-08
	173.0-176.2	0.938	0.501	0.103	0.700	0.0340	9.08E-09
	189.8-193.0	0.594	0.453	1.000	0.700	0.1883	1.01E-06
	196.6-199.8	0.563	0.434	0.500	1.000	0.1220	6.00E-08
	213.2-216.0	0.679	0.387	1.000	1.000	0.2625	4.54E-07
	249.0-251.8	0.393	0.285	0.091	0.700	0.0071	4.03E-09
	272.0-274.8	0.214	0.219	1.000	0.700	0.0328	3.36E-08

It should be noted that the values of (1-GCD) in HB-94-01 borehole are all equal to 1. The results of Eq. 6 demonstrate that the empirical HC model may also be more accurate for the estimation of the rock mass hydraulic conductivity if the fractures do not contain infillings. There are a few limitations that need to be noted for the use of Eq. 5. The data used to develop the equation are limited in number and in the lithologies represented. From the definition of DI, DI cannot be determined for inclined boreholes because the data collected were from vertical boreholes.

Model Verification

In order to further verify the feasibility of the proposed empirical HC model, the model verification is conducted. Another borehole data with the drilling depth of 120 m is adopted to verify the empirical HC model. The principal lithologic units of the borehole, namely CH- 04, are mainly sandstone, shale, and sandstone with some thin shale. The depth from 24.5 m to 26.6 m, 32.5 m to 34.1 m, 65.7 m to 67.8 m, and 77.8 m to 79.9 m were sealed by double packers for conducting the hydraulic tests. The quantitative evaluation of hydraulic parameters was then performed using AQTESOLVE which uses an iterative process of the best-fit theoretical response curves based on the measured data of the hydraulic test. Figure 8 shows that the comparison of the rock mass hydraulic conductivity obtained by in-situ test and that from the estimation of the empirical HC model. Very good correlation can be found (Fig. 8). This verification example demonstrates that the empirical HC model is able to determine the rock mass hydraulic conductivity for different sites in which the lithologic conditions are similar.

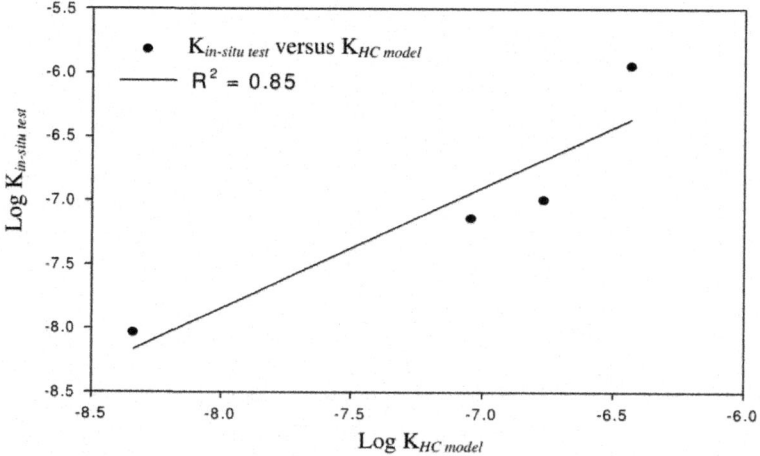

Figure 8. Correlation between Kin-situ and KHC model.

Table 4. Four hydraulic test data for the model verification (Borehole CH-04).

Test intervals (m)	RQD(%)	DI	1-GCD	LPI	HC	$K_{HC\ model}$	$K_{in\text{-}situ}$
24.5~26.6	81.0	0.787	0.952	0.55	0.0785	9.06E-08	7.14E-08
32.5~34.1	43.8	0.723	0.975	0.55	0.2179	3.69E-07	1.11E-06
65.7~67.8	47.6	0.444	0.976	0.55	0.1248	1.71E-07	9.95E-08
77.8~79.9	95.2	0.343	1.000	0.55	0.0090	4.59E-09	9.09E-09

$K_{HC\ model}$ and $K_{in\text{-}situ}$ represent K obtained by Eq. 5 and the in-situ hydraulic test, respectively.

CONCLUSIONS

The estimation of rock mass hydraulic conductivity of highly disturbed clastic sedimentary rocks in Taiwan was performed using the data of BHTV and double packer hydraulic tests. The field results indicated that the rock mass in the study area has the conductivity between the order 10-10 and 10-6 m/s at the depth between 34 m and 275 m below ground surface. The results demonstrate that the rock mass hydraulic conductivity of highly disturbed clastic sedimentary rocks in Taiwan mainly depends on the following four parameters: RQD, DI, GCD, and LPI.

This paper proposes an empirical HC model for estimating rock mass hydraulic using data collected for highly disturbed clastic sedimentary rocks in Taiwan. The HC-value can be calculated from borehole image data and rock core data. To verify rationality of the proposed HC model, the study collected data from the results of two hydrogeological investigation programs in three boreholes to determine a relationship between hydraulic conductivity and HC. Besides, good correlation is found from the model verification which demonstrates that the empirical HC model is able to determine the rock mass hydraulic conductivity for different sites in which the lithologic conditions are similar. The regression results indicated that the relationship of a power law exists between the two variables with a coefficient of determination of 0.866. The empirical HC model may provide a useful tool to predict hydraulic conductivity of fractured rocks based on measured HCvalues. By using this model, hydraulic conductivity data in a given site can be directly acquired, which removes the cost on hydraulic tests. For in-situ aquifer tests, the empirical HC model is valuable for preliminary assessment of the degree of permeability in a packedoff interval of a borehole.

REFERENCES

1.   Bear, J. (1972). Dynamics of Fluids in Porous Materials, American Elsevier.

2.   Black, J. H. (1987). Flow and flow mechanisms in crystalline rock, in

Fluid Flow in Sedimentary Basins and Aquifers. Geol. Soc. Special Publication No. 34, 186-200.

3.   Burgess, A. (1977). Groundwater Movements Around a Repository— Regional Groundwater Analysis. Kaernbraenslesaekerhet, Stockholm, Sweden, 116.

4.   Carlson, A. & Olsson, T. (1977). Hydraulic properties of Swedish crystalline rocks-hydraulic conductivity and its relation to depth. Bulletin of the Geological Institute, University of Uppsala 7, 71-84.

5.   Deere, D. U.; Hendron, A. J.; Patton, F. D.& Cording, E. J. (1967). Design of surface and near surface construction in rock. Proceedings of 8th U.S. Symposium Rock Mechanics, AIME, New York 237-302.

6.   Duffield, G. M. (2004). AQTESOLVE version 4 user's guide, Developer of AQTESOLV HydroDOLVE, Inc., Reston, VA, USA.

7.   National Research Council. (1996). Rock fractures and fluid flow: contemporary understanding and applications. National Academy Press, Washington D. C., USA.

8.   Singhal, B. B. S. & Gupta, R. P.(1999). Applied hydrogeology of fractured rocks. Kluwer Academic Publishers, The Netherlands, 400.

9.   Spitz, K. & Morena, J. (1996). A Practical Guide to Groundwater and Solute Transport Modeling, Wiley.

10.   Sinotech Engineering Consultants, LTD. (2006). Tseng-Wen transbasin diversion tunnel project-supplemental geology investigation, Southern Water Resources Office, Water Resources Agency, Ministry of Economic Affairs, Taiwan (in Chinese).

11.   Snow, D. T. (1969). Anisotropic permeability of fractured media. Water Resources Research, 5(6), 1273-1289.

12.   Wei, Z.Q., Egger, P., Descoeudres, F. (1995). Permeability predictions for jointed rock masses. International Journal of Rock Mechanics, Mineral Science and Geomechanics 32, 251-261.

13.   Williams, J. H. & Johnson, C. D. (2004). Acoustic and optical borehole-wall imaging for fractured-rock aquifer studies. Journal of Applied Geophysics, 55(1–2): 151–159.

14.   Zhao, J. (1998). Rock mass hydraulic conductivity of the Bukit Timah granite, Singapor. Engineering Geology, V 50, 211-216.

Chapter 8

EFFECT OF PARTIAL WATER SATURATION ON ATTENUATION CHARACTERISTICS OF LOW POROSITY ROCKS

Tae-Min Oh[1], Tae-Hyuk Kwon[2], Gye-Chun Cho[1]

[1] Department of Civil and Environmental Engineering, Korea Advanced Institute of Science and Technology (KAIST), Taejon 305-701, Korea

[2] Earth Sciences Division, Lawrence Berkeley National Laboratory, 1 Cyclotron Rd. MS 90R1116, Berkeley CA 94720, USA

INTRODUCTION

With the advance of geophysical exploration techniques, in situ measurements on elastic wave velocities have been successfully employed for investigating near-surface and deep geological structures. However, in situ techniques that measure the attenuation of elastic waves are still in their infancy due to the difficulty of calibration, the lack of theoretical models, and the inaccuracy of field scale measurements. While attenuation itself is strongly related to important physical characteristics of porous media, such as pore fluid composition, stress states, and internal heterogeneity (Johnston et al. 1979), the understanding of the attenuation characteristics of porous media is still limited.

The rocks are frequently exposed to water through rain or ground water. The presence of pore water not only plays an important role from an engineering perspective, but also significantly alters acoustic wave propagation. Compressional wave velocity changes little until the pore spaces are fully saturated with water because the air in partially saturated pore fluids diminishes the stiffness of the pore fluids and hardly contributes to strengthening the rock frame. On the other hand, when there is an increase in water saturation, the compressional wave attenuation—that is related to energy dissipation—tends to increase more sensitively than velocity (Gardner et al. 1964; Toksöz et al. 1979; Mavko and Nur 1979; Murphy 1982; Winkler and Nur 1982; Cadoret et al. 1998).

Several previous researchers have investigated the fluid effect on the wave velocity and attenuation, mainly focusing on high porosity rocks (mostly greater than 20% of porosity), such as limestone (Cadoret et al.1998) and sandstone (Murphy 1982; Winkler and Nur 1982). This study, therefore, examines how water saturation affects the attenuation characteristics of low porosity rocks. The effect of partial water saturation on the attenuation is explored by recapitulating Biot model and by conducting a series of laboratory tests. Then, the results of Biot model and laboratory tests are compared and some implications are discussed.

RECAPITULATION OF BIOT MODEL: PARTIAL WATER SATURATION EFFECT ON ATTENUATION

Biot model (Biot 1956a, b) can be applied to porous media, particularly for modeling wave velocity and attenuation of a rock. Biot model assumes that pore fluids flow only in the direction parallel to the direction of the wave propagation and that the viscous fluid motion follows Poiseuille flow in a cylindrical tube. No chemical, electrical, or thermal interactions between different phases are taken into account in the model. Detailed formulations of the Biot model and the physical parameters used are summarized in Appendix A. The following parametric study recapitulates Biot model and explores how various physical properties of rocks, such as porosity, permeability, or pore size parameter, affect the attenuation of compressional wave (hereafter, P-wave) in a partially water-saturated porous medium. The pore fluid properties (e.g., bulk modulus and density), in which the composition of water and air vary, are calculated by the averaging method (more details are described in the Appendix section).

Porosity

Attenuation increases as the porosity increases because the water content in a pore becomes larger at the same degree of water saturation. Higher porosity rocks can have more opportunity of interaction between fluid and rock frame in the cracks than lower porosity rocks. Specifically, Figure. 1a shows that the attenuation of a high porosity rock (when $n = 30\%$, $Q^{-1} = 0.0046$) is approximately nine times larger than that of a low porosity rock (when $n = 1\%$, $Q^{-1} = 0.0005$) at a saturation degree of 98–99%, which is the saturation degree required for a maximum attenuation value.

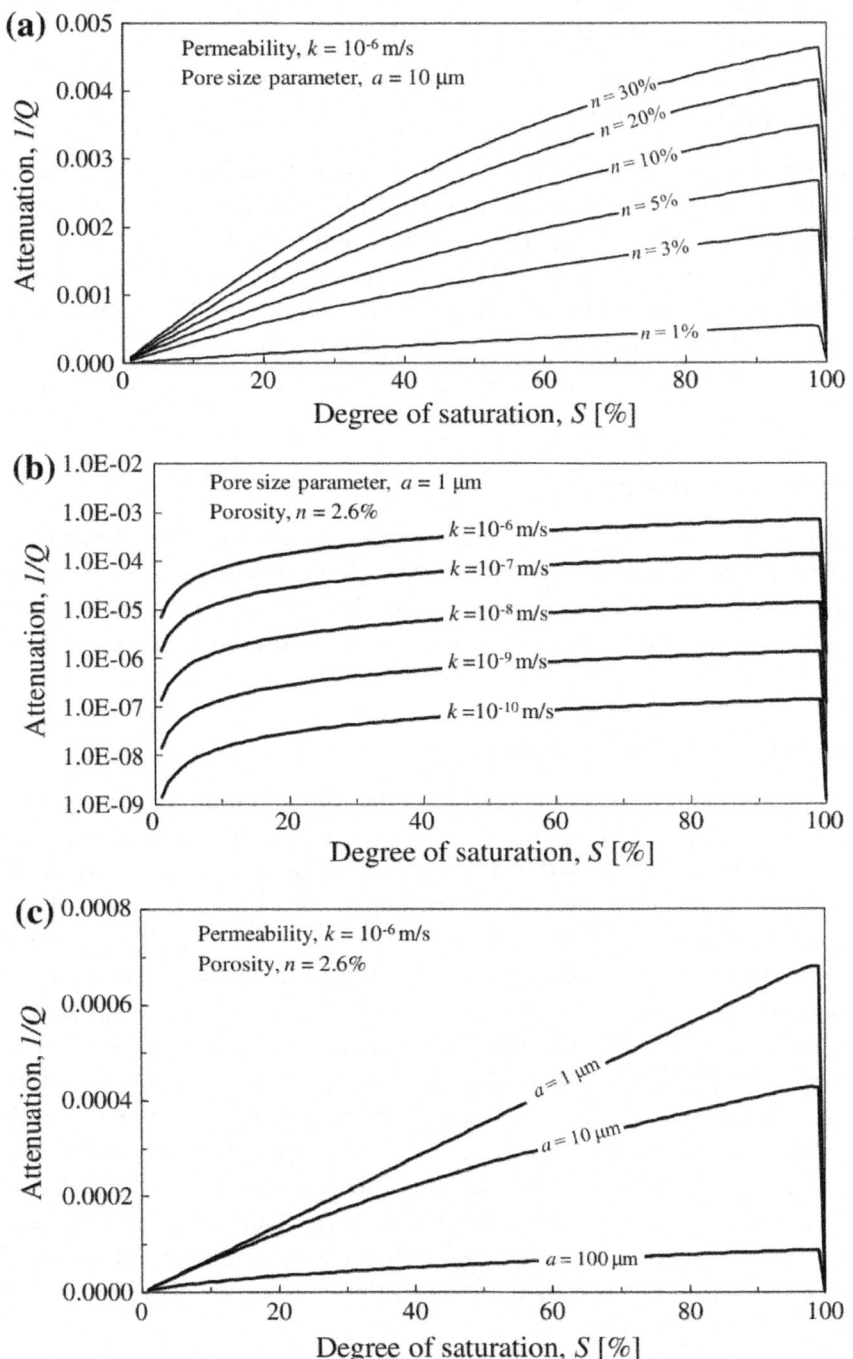

Figure. 1: Attenuation versus water saturation varying **a** porosity, **b** permeability,

and **c** pore size parameter when the Young's modulus E = 45 GPa, frequency f = 23 kHz, and Poisson's ratio v = 0.24.

Permeability

For low permeability rocks, the pressure gradient induced by wave propagation dissipates slowly and there are minimal flows of fluid in the pores and pore throats. Meanwhile, for higher permeability rocks, fluids can move more freely, and consequently cause greater loss of energy during the wave propagation. It is found that permeability has the strongest impact on the magnitude of the attenuation than porosity and pore size, affecting approximately one order of magnitude increase of attenuation by an order of magnitude increase of the permeability, as shown in Figure. 1b.

Pore Size Parameter

The pore size parameter is an indicator of the geometric characteristics of the pore space. As shown in Figure. 1c, when the rock has the same porosity and permeability, the attenuation increases as the pore size parameter decreases. It implies that the wave loses more energy in smaller-sized pores at the same porosity.

Attenuation Curve Tendency

The attenuation (Q^{-1}) is normalized by the maximum attenuation (Q_{max}^{-1}), which is the attenuation value for nearly full water-saturated conditions (i.e., S ≈ 99%). The normalized value enables us to examine how the curve tendency and shape depend on the various parameters of low porosity rocks. The attenuation-water saturation (Q^{-1}/Q_{max}^{-1} − S) curves (hereafter, attenuation curve) are superimposed in Figure. 2 to highlight the effect of changes in the permeability and pore size parameter. Porosity is excluded as a variable because it has a negligible effect on the attenuation curve tendency. Figure 2 shows that the attenuation curve shape with water saturation can be convex, linear, or concave depending on the permeability. A high permeability ($k ≥ 10^{-5}$ m/s in the presented study) produces a concave attenuation curve. In contrast, the attenuation curve is convex when the permeability is lower than 10^{-5} m/s. It is worth to note that a small pore size parameter produces a linear curve: the attenuation curve is almost linear with water saturation when the pore size parameter is less than 1 μm (also refer to Figure. 1c).

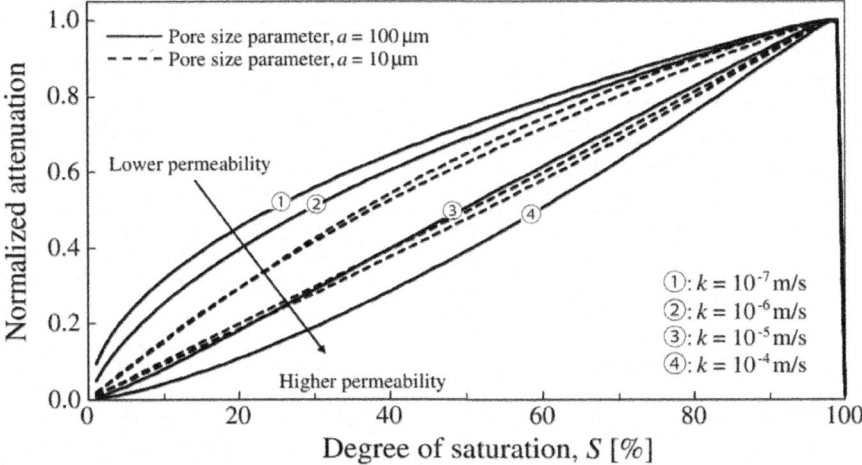

Figure. 2: Normalized attenuation curves (Q^{-1}/Q_{max}^{-1}) with respect to a range of the permeabilities and pore sizes when the porosity $n = 2.6\%$, Young's modulus $E = 45$ GPa, frequency $f = 23$ kHz, and Poisson's ratio$v = 0.24$.

EXPERIMENTAL PROGRAM

Specimen Preparation

Four cylindrical rock specimens were prepared, representing unique low porosity characteristics of deep bed-rock formations in Korea: two weathered granite and two mudstone specimens. All the specimens had a diameter of 5 cm, a length of 10 cm, and a porosity value of less than 3%. The dry density of the specimens was measured after drying them in an oven at 50°C for 72 h, while the water-saturated density was measured after submerging them until the measured masses converged to a certain value. The effective porosity, which is a measure of connected pores, but not closed pores, was estimated from a dry density and a fully water-saturated density. A standard method suggested by the International Society for Rock Mechanics (ISRM; Brown 1981) was used to determine the specific gravity of each specimen. The properties of the specimens tested are summarized in Table 1.

Table 1: Properties of tested specimens

Specimen	Granite 1	Granite 2	Mudstone 1	Mudstone 2
Symbol	GN1	GN2	MS1	MS2
Density (kg/m³)				

Dry	2,591	2,596	2,628	2,626
Saturation	2,603	2,609	2,637	2,652
Porosity (%)	1.16	1.26	0.96	2.61
Specific gravity (–)	2.62	2.62	2.65	2.64
Rod-wave velocity (m/s)[a]	1,864	1,105	4,891	4,137
P-wave velocity (m/s)[b]	2,619	1,625	5,273	4,459
Young's modulus (GPa)[c]	9.4	3.3	62.9	45
Poisson's ratio (–)[d]	0.38	0.39	0.24	0.24

[a]Rod-wave velocity of a dry specimen at the resonance frequency was determined by using a free–free resonant column (FFRC) method (Vaghela and Stokoe 1995; Kim et al. 1997; Cha and Cho 2007)

[b]P-wave velocity of a dry specimen was determined by using the point-source travel time method (Cha and Cho 2007)

[c]Young's modulus (i.e., $E = \rho V_p^2$) was calculated by using the dry density and the P-wave velocity of a specimen at a dry condition

[d]Poisson's ratio is calculated by using the P-wave velocity and rod-wave velocity of a specimen at a dry condition

Saturating and Drying Processes

To determine water content, porosity, density, absorption, and related properties, saturating and drying processes were performed by following the procedures suggested by ISRM (Brown 1981). A saturating process was performed by completely immersing a specimen in a water bath. The water bath was agitated to remove any trapped air. After a saturating procedure, the specimen was dried in an oven at a constant temperature of 50°C in order to prevent any damage induced by thermal shock. As time elapses, the degree of water saturation was periodically evaluated by measuring the mass, and the attenuation was simultaneously measured.

Free–Free Resonant Column Test

The attenuation of a rock specimen was obtained by a free–free resonant column (FFRC) test (Vaghela and Stokoe 1995; Kim et al. 1997; Cha and Cho 2007). The FFRC method measures the attenuation of the longitudinal wave mode (i.e., also called as rod wave or bar wave). A typical result of the FFRC test at the time domain is shown in Figure. 3a. The damping ratio

(D) is obtained by fitting the frequency response curve of a single degree of freedom system to the measured data as shown in Figure. 3b. The attenuation is expressed in terms of the inverse quality factor (Q^{-1}) calculated by the relationship, $Q^{-1} = 2D$.

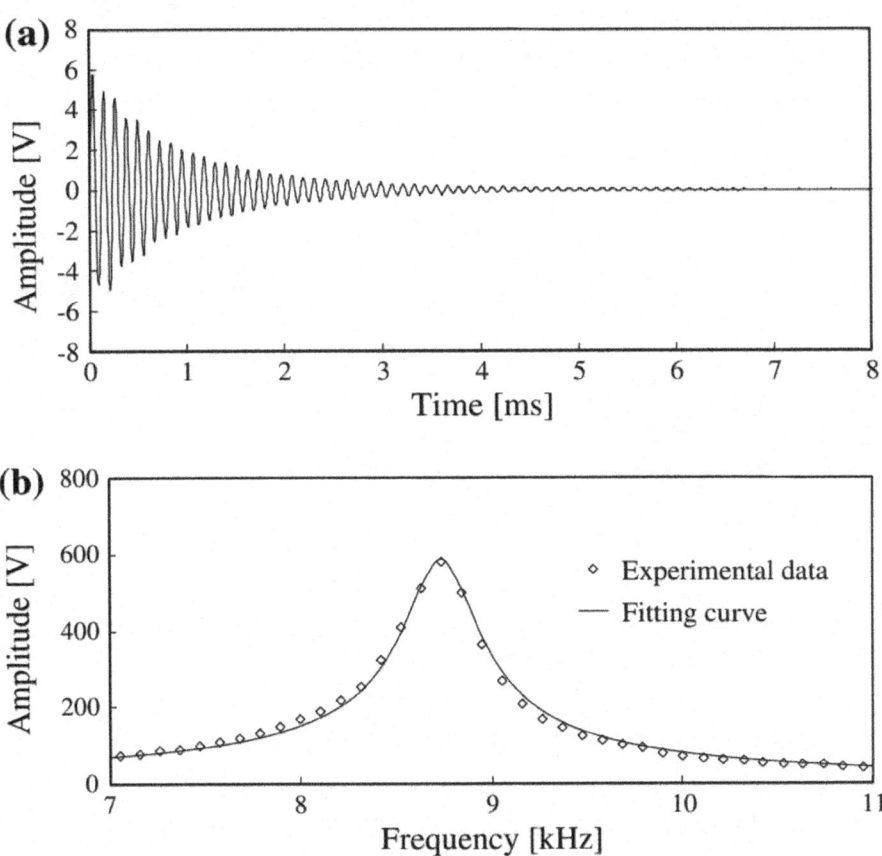

Figure. 3: A typical signal measured by an FFRC test at **a** the time domain and **b** the frequency domain.

Attenuation is highly affected by frequency and confining pressure. All the measurements were taken under atmospheric pressure and a room temperature of 25°C. The FFRC measurements were conducted within the regime of the resonant frequency between 6 and 27.5 kHz. The signals were captured regularly until the measured mass of the specimen converged to a certain value, where the specimen was considered as fully water-saturated. We subsequently subjected the saturated specimens to a drying process in an oven and repeated the same elastic wave and mass measurement procedure.

RESULTS AND DISCUSSION

Overall, the experimental results deviate far from Biot model's prediction, showing that the experimental attenuation results (Figure. 4) are approximately orders of magnitude greater than Biot attenuation results (Figure. 1). This underestimation may be due to the fact that Biot model does not take random micro-cracks which are closely related to permeability into consideration.

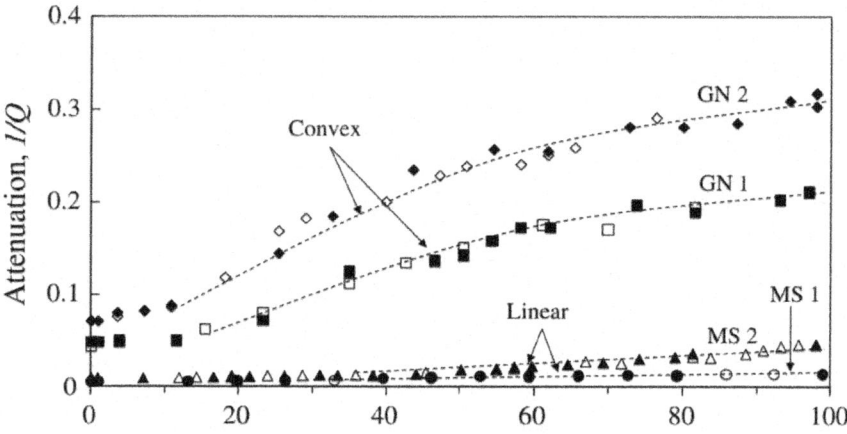

Figure. 4: Attenuations of longitudinal waves at difference water saturations. Note that the open points indicate the measurements during drying process whereas the solid points indicate the measurements during saturating process.

Attenuation under a Dry Condition

According to Biot model, the wave does not dissipate under a dry condition (i.e., 0% saturation); hence, the attenuation is expected to be null ($Q^{-1} \to 0$ when $S \to 0$). However, attenuations are unexceptionally measured for natural rock specimens under a dry condition (e.g., GN1: 0.048, GN2: 0.071, MS1: 0.006, and MS2: 0.010). These experimental measurements represent pure material damping without fluid effect, where the material damping is decided by frictional energy loss between cracks (or mineral grain boundaries) during elastic wave propagation (Winkler and Nur 1982). As shown in Figure. 4, granite specimens have higher attenuations under a dry condition than the mudstone specimens do because the mudstone specimens are much stiffer and have a smaller particle size than the granite specimens. It indicates that the energy loss due to internal heterogeneity-induced wave scattering, such as micro-cracks, is greater in the granite specimens than in the mudstone specimens.

Attenuation under Partial Water Saturation

Attenuation curve tendency can be considered to have two distinctive regimes: low saturation regime and mid-to-high saturation regime. In the low saturation regime, the attenuation behavior as a function of saturation is dominated by the microscopic fluid flow mechanism (e.g., as explained by the squirt flow model; Cadoret et al. 1998). On the other hand, attenuation behavior in the mid-to-high saturation range is governed by a macroscopic mechanism (e.g., as explained by Biot model).

In a low water saturation regime, as shown in Figure. 4, attenuation of the low porosity rocks hardly or slightly increases with water saturation. This region covers the saturation from 0 to approximately 10 or 20%. In contrast, in high-porosity rocks (e.g., Murphy 1982; Cadoret et al. 1998), the region that provides no or slight increase in attenuation with increasing water saturation covers a wider range of water saturation, i.e., from 0 to approximately 60 or 70%. This phenomenon is due to the fact that the high porosity rock requires more water to reveal Biot effect, which causes water–solid coupling and decoupling.

In the mid-to-high water saturation regime, Biot model can be applicable for predicting the attenuation tendency and comparing it to measurements because Biot model is based on a macroscopic mechanism. Typical attenuation-water saturation curves are shaped as convex, linear, or concave (see Figure. 2). The attenuation curves of the granite specimens are convex, while the curves of the mudstone specimens are linear in the mid-to-high water saturation regime, as shown in Figure. 4. The linearity of the attenuation curves of the mudstone specimens is attributed to the small pore size of the mudstone specimen. In contrast to low porosity rocks, it has been presented in previous studies (e.g., Murphy 1982; Cadoret et al. 1998) that the attenuation curves of high-porosity rocks (n > 23%) are presumed to be concave in the mid-to-high saturation regime. This concave shape of the attenuation curve in high-porosity rocks is consistent with the calculations in Biot model, as shown in Figure. 2 (i.e., when permeability k is larger than 10^{-5} m/s and the pore size parameter a is larger than 10 μm).

A hysterical loop of attenuation during the saturating and drying processes is generally evident in tests on high porosity rocks due to different water distribution conditions (Cadoret et al. 1998). In contrast, the fluid distribution during the saturating and drying processes of low porosity rock has little effect on the attenuation. Figure 4 shows that the attenuation values during the saturation process are almost the same as those of the drying process for low porosity rocks.

Attenuation under a Fully Water-Saturated Condition

The attenuation values of the tested specimens when fully saturated are nearly two to five times greater than the values under a dry condition as shown in Table 2 and Figure. 4. In particular, Figure. 4 indicates that the granite specimens are strongly influenced by a saturation level. GN2 (the softest rock in the presented study) has larger difference in attenuation between dry and saturation conditions than does MS1 (the hardest rock in the presented study). This means that a soft rock (e.g., a rock with low stiffness and slow wave velocity) is easily affected by fluid flow during saturation.

Table 2: Measured wave velocities and attenuations of the tested specimens

Specimen	Resonant frequency (kHz)	P-wave velocity (m/s)		Rod-wave velocity (m/s)		Attenuation ($1/Q$) (–)	
		Dry	Satu-rated	Dry	Saturated	Dry (Min.)	Saturated (Max.)
Granite 1 (GN1)	9.5	2,619	2,794	1,864	1,906	0.048	0.210
Granite 2 (GN2)	6.0	1,625	1,820	1,105	1,275	0.071	0.316
Mudstone 1 (MS1)	27.5	5,273	5,318	4,891	4,966	0.006	0.014
Mudstone 2 (MS2)	23.0	4,459	4,508	4,137	4,175	0.010	0.046

It is worth noting that high porosity rocks demonstrate attenuation drops in a fully water saturated condition (refer to the data in Murphy 1982; Cadoret et al. 1998). This drop is because no patchy water distribution and no squirt flow occur when water fully saturates the pores in a rock; as a result, the pore fluid cannot easily squeeze or flow into adjacent pores. Thus, the energy loss decreases. However, this phenomenon is not observed in the low porosity rock specimens tested (refer to Figure. 4). This result implies that low porosity rocks are unlikely to be fully water-saturated under a natural atmospheric environment without some additional pressure due to water surface tension on the micro-cracks and closed pores, both of which form during rock diagenesis. This condition is expected in various near-surface grounds.

CONCLUSIONS

This study explored how water saturation affects the compressional wave attenuation characteristics in low porosity rocks by reviewing Biot model and performing a series of experiments. Main findings are as follows:

- Biot model shows that permeability is the most influential factor on attenuation due to wave energy dissipation by fluid flow between cracks (or pores).

- The experimental results deviate far from Biot model's prediction, showing that the experimental attenuation results are approximately orders of magnitude greater than Biot attenuation results. This discrepancy (i.e., Biot model's underestimation) may be due to the fact that Biot model does not take into consideration random micro-cracks which are closely related to permeability.

- The curve of attenuation versus saturation can be convex, linear, or concave from low to high permeability. The experimental results show that the attenuation of a low porosity rock tends to be convex or linear and shows no hysterical loop during the saturating and drying process.

- In the low saturation regime, the attenuation of low porosity rocks is dominated by the microscopic fluid flow mechanism (e.g., Squirt flow model) while in the mid-to-high saturation range, it is governed by a macroscopic mechanism (e.g., Biot model).

ACKNOWLEDGMENTS

This work was supported by the Smart Infra-Structures Technology Center (SISTeC) under KOSEF and the grant (07UrbanRenaissanceB03) from High-Tech Urban Development Program funded by the Ministry of Construction & Transportation of Korean Government.

APPENDIX A. BIOT MODEL FOR POROUS MEDIA

Fast P-wave velocity (V_p) is calculated by the following formulations:

$$\left(\frac{H}{V_p^{*2}} - \rho_{mix}\right) \cdot \left(q - \frac{I}{V_p^{*2}}\right) - \left(\frac{C}{V_p^{*2}} - \rho_f\right) \cdot \left(\rho_f - \frac{C}{V_p^{*2}}\right)$$
$$= 0, \text{ and} \tag{1}$$

$$V_p = \text{Re}(V_p^*), \tag{2}$$

where V_p^* is the complex P-wave velocity, and $\text{Re}(V_p^*)$ is the real part of the complex P-wave velocity. The parameters A, I, H, C, F, and q are defined as follows:

$$A = B_g \cdot \left(1 + n \cdot \left(\frac{B_g}{B_f} - 1\right)\right), \tag{3a}$$

$$I = \frac{B_g^2}{A - B_{sk}},$$

(3b)

$$H = B_{sk} + \frac{4}{3} \cdot G_{sk} + \frac{(B_g - B_{sk})^2}{A - B_{sk}},$$

(3c)

$$C = \frac{B_g - B_{sk}}{A - B_{sk}} \cdot B_g,$$

(3d)

$$F = \frac{\zeta \cdot T}{4 \cdot \left(1 + \frac{2 \cdot j \cdot T}{\zeta}\right)}, \text{ and}$$

(3e)

$$q = \frac{\alpha \cdot \rho_f}{n} - j \cdot \frac{\eta \cdot F}{\omega \cdot K}.$$

(3f)

Attenuation (1/Q) is expressed as follows:

$$\frac{1}{Q} = \frac{\text{Im}(M^*)}{\text{Re}(M^*)},$$

(4)

where Q is the quality factor, and M^* is the complex constraint modulus. $\text{Im}(M^*)$ means the imaginary part, and $\text{Re}(M^*)$ describes the real part of the complex constraint modulus (Table 3).

Table 3: Parameters used in this study

Parameter	Definition	Values selected for this study
a	Pore size parameter	
B_a	Air bulk modulus	142 kPa at 1 atm[a]
B_f	Fluid bulk modulus	$B_f = \frac{1}{\frac{S}{B_w} - \frac{1-S}{B_a}}$
B_g	Grain (particle) bulk modulus	50 GPa[a]
B_{sk}	Skeleton bulk modulus	$B_{sk} = E/3(1 - 2v)$
B_w	Water bulk modulus	2.18 GPa at 1 atm[a]
E	Elastic modulus	Dynamic Young's modulus
G_{sk}	Skeleton shear modulus	$G_{sk} = E/2(1 + v)$
k	Permeability	
K	Absolute hydraulic conductivity	$K = \eta \cdot k/(g \cdot \rho_f)$ where g is gravitational acceleration ($g = 9.8$ m/s^2)

n	Porosity	
S	Degree of saturation	
T	Visco-dynamic operator	$T = e^{(3/4)j\pi}(J_1(\zeta \cdot e^{-j \cdot \pi/4})/ J_0(\zeta \cdot e^{-j \cdot \pi/4}))^b$
		J_1 and J_0 are Bessel functions and $j^2 = -1$
α	Tortuosity factor	$\alpha = 0.8(1 - n) + 1^c$
		$\alpha = 1$ for tubes, $\alpha = 2$–3 for particulate materials
ρ	Mass density	$\rho_g = 2,650$ kg/m^3 of grain(particle) density
		$\rho_w = 1,000$ kg/m^3 of water
		$\rho_f = S\rho_w$ of pore fluid
		$\rho_{mix} = (1 - n)\rho_g + n \cdot S \cdot \rho_w$ of mixture composed of porous medium with fluid
η	Dynamic viscosity	0.001 Pa s at 20°C for water
ζ	Dimensionless factor	$\zeta = a(\omega \cdot \rho_f/\eta)^{0.5b}$
v	Poisson's ratio	
ω	Angular frequency	$\omega = 2\pi f$ (used resonant frequency for f)

[a]Santamarina et al. (2001)

[b]Stoll (1980)

[c]Koponen et al. (1996)

REFERENCES

1. Biot MA (1956a) Theory of propagation of elastic wave in a fluid-saturated porous solid: I. Low-frequency range. J Acoust Soc Am 28(2):168–178

2. Biot MA (1956b) Theory of propagation of elastic wave in a fluid-saturated porous solid: II. Higher frequency range. J Acoust Soc Am 28(2):179–191

3. Brown ET (1981) Rock characterization testing & monitoring. Pergamon Press, Oxford

4. Cadoret T, Mavko G, Zinszner B (1998) Fluid distribution effect on sonic attenuation in partially saturated limestones. Geophysics 63(1):154–160

5. Cha MS, Cho GC (2007) Compression wave velocity of cylindrical rock specimens: engineering modulus interpretation. Jpn J Appl Phys 46(7B):4497–4499

6. Gardner GHF, Wyllie MRJ, Droschak DM (1964) Effects of pressure and

fluid saturation on the attenuation of elastic waves in sands. J Petr Tech 16(2):189–198

7. Johnston DH, Toksőz MN, Timur A (1979) Attenuation of seismic waves in dry and saturated rocks: II. Mechanisms. Geophysics 44(4):691–711

8. Kim DS, Kweon GC, Lee KH (1997) Alternative method of determining resilient modulus of compacted subgrade soils using free-free resonant column test. Transport Res Rec 1577:62–69

9. Koponen A, Kataja M, Timonen J (1996) Tortuous flow in porous media. Phys Rev E 54(1):406–410

10. Mavko GM, Nur A (1979) Wave attenuation in partially saturated rocks. Geophysics 44(2):161–178

11. Murphy WF (1982) Effects of partial water saturation on attenuation in Massilon sandstone and Vycor porous glass. J Acoust Soc Am 71(6):1458–1467

12. Santamarina JC, Klein KA, Fam MA (2001) Soils and waves. Wiley, New York

13. Stoll RD (1980) Theoretical aspects of sound transmission in sediments. J Acoust Soc Am 68(5):1341–1349

14. Toksőz MN, Johnston DH, Timur A (1979) Attenuation of seismic waves in dry and saturated rocks: I. Laboratory measurements. Geophysics 44(4):681–690

15. Vaghela JG, Stokoe KH (1995) Small-strain dynamic properties of dry sand from the free-free resonant column. Geotechnical engineering report GT95-1, Geotechnical Engineering Center, University of Texas at Austin, Austin

16. Winkler KW, Nur A (1982) Seismic attenuation: effects of pore fluids and frictional sliding. Geophysics 46(1):1–15

CITATION

CHAPTER 1

Meng Wang, Zheming Zhu, and Jun Xie, "An Experimental Study on Deformation Fractures of Fissured Rock around Tunnels in True Triaxial Unloads," Advances in Materials Science and Engineering, vol. 2015, Article ID 982842, 10 pages, 2015. doi:10.1155/2015/982842.

CHAPTER 2

Yifeng Chen and Chuangbing Zhou (2011). Stress/Strain-Dependent Properties of Hydraulic Conductivity for Fractured Rocks, Developments in Hydraulic Conductivity Research, Dr. Oagile Dikinya (Ed.), ISBN: 978-953-307-470-2, InTech, DOI: 10.5772/16007.

CHAPTER 3

Are Ha°vard Høien and Bjørn Nilsen, "Rock Mass Grouting in the Løren Tunnel: Case Study with the Main Focus on the Groutability and Feasibility of Drill Parameter Interpretation," Rock Mech Rock Eng (2014) 47:967–983 DOI 10.1007/s00603-013-0386-7.

CHAPTER 4

Diyuan Li, Charlie C. Li and Xibing Li, "Influence of Sample Height-to-Width Ratios on Failure Mode for Rectangular Prism Samples of Hard Rock Loaded In Uniaxial Compression," Rock Mech Rock Eng (2011) 44:253–267 DOI 10.1007/s00603-010-0127-0.

CHAPTER 5

S. R. Hencher, S. G. Lee, T. G. Carter and L. R. Richards, "Sheeting Joints: Characterisation, Shear Strength and Engineering," Rock Mech Rock Eng (2011) 44:1–22 DOI 10.1007/s00603-010-0100-y.

CHAPTER 6

Takahiro Funatsu, Norikazu Shimizu, Mahinda Kuruppu and Kikuo Matsui, "Evaluation of Mode I Fracture Toughness Assisted by the Numerical Determination of K-Resistance," Rock Mech Rock Eng (2015) 48:143–157 DOI 10.1007/s00603-014-0550-8.

CHAPTER 7

Cheng-Yu Ku and Shih-Meng Hsu (2011). Estimating Hydraulic Conductivity of Highly Disturbed Clastic Rocks in Taiwan, Hydraulic Conductivity - Issues, Determination and Applications, Prof. Lakshmanan Elango (Ed.), ISBN: 978-953-307-288-3, InTech, DOI: 10.5772/18553.

CHAPTER 8

Tae-Min Oh, Tae-Hyuk Kwon and Gye-Chun Cho, "Effect of Partial Water Saturation on Attenuation Characteristics of Low Porosity Rocks," Rock Mech Rock Eng (2011) 44:245–251 DOI 10.1007/s00603-010-0121-6.

INDEX

A

Atomic Energy of Canada Limited (AECL's) 152

B

Brazilian disk test (BDT) 210

C

chevron bend (CB) 209, 210, 216, 223, 233

chevron-notched Brazilian disk (CC-NBD) 210

completely decomposed granite (CDG) 163

cracked straight-through Brazilian disk (CSTBD) 210

crack initiation (CI) 125

crack-opening displacement (COD) 214, 215, 234

cubic block 75, 76, 77, 78, 80

D

depth index (DI) 240, 251

distinct element method (DEM) 209, 212, 220

drill parameter interpretation (DPI) 87, 89, 90, 103

E

elasticity 3

elastic wave velocities 259

excavation 87, 90, 91, 95, 96, 106, 110, 115

F

finite element method (FEM) 212

fracture process zone (FPZ) 211, 233

free–free resonant column (FFRC) 264

G

gouge content designation (GCD) 240, 251

Gouge Content Designation (GCD) 253

H

highly decomposed granite (HDG) 163

hydraulic aperture 26, 29, 34, 36, 44, 45

hydraulic conductivity 25, 26, 27, 28, 29, 30, 31, 34, 35, 36, 39, 41, 43, 44, 45, 49, 51, 52, 53, 54, 55, 56, 57, 59, 61, 62, 67, 68, 70, 71, 74, 75, 76, 77, 78, 79, 80, 81, 82, 83, 84, 85